Introduction to
Stochastic Processes

World Scientific Series on Probability Theory and Its Applications

Print ISSN: 2737-4467
Online ISSN: 2737-4475

Series Editors: Zenghu Li *(Beijing Normal University, China)*
Yimin Xiao *(Michigan State University, USA)*

Published:

Vol. 2   *Introduction to Stochastic Processes*
by Mu-Fa Chen (Beijing Normal University, China) and
Yong-Hua Mao (Beijing Normal University, China)

Vol. 1   *Random Matrices and Random Partitions: Normal Convergence*
by Zhonggen Su (Zhejiang University, China)

World Scientific Series on
**Probability Theory and
Its Applications**

Volume 2

Introduction to
Stochastic Processes

Mu-Fa Chen
Yong-Hua Mao

Beijing Normal University, China

**Higher
Education
Press**

World Scientific

NEW JERSEY · LONDON · SINGAPORE · BEIJING · SHANGHAI · HONG KONG · TAIPEI · CHENNAI · TOKYO

Published by

Higher Education Press Limited Company
4 Dewai Dajie, Beijing 100120, P. R. China
and
World Scientific Publishing Co. Pte. Ltd.
5 Toh Tuck Link, Singapore 596224
USA office: 27 Warren Street, Suite 401-402, Hackensack, NJ 07601
UK office: 57 Shelton Street, Covent Garden, London WC2H 9HE

Library of Congress Cataloging-in-Publication Data
Names: Chen, Mufa, author. | Mao, Yong-Hua, author.
Title: Introduction to stochastic processes / Mu-Fa Chen, Yong-Hua Mao,
    Beijing Normal University, China.
Description: New Jersey : World Scientific, [2021] | Series: World scientific series on
    probability theory and its applications, 2737-4467 ; Vol. 2 |
    Includes bibliographical references and index.
Identifiers: LCCN 2020044704 | ISBN 9789814740302 (hardcover) |
    ISBN 9789814740319 (ebook for institutions) | ISBN 9789814740326 (ebook for individuals)
Subjects: LCSH: Stochastic processes.
Classification: LCC QA274 .C446 2021 | DDC 519.2/3--dc23
LC record available at https://lccn.loc.gov/2020044704

British Library Cataloguing-in-Publication Data
A catalogue record for this book is available from the British Library.

For any available supplementary material, please visit
https://www.worldscientific.com/worldscibooks/10.1142/9903#t=suppl

Preface to the English Edition

This is an English edition of our Chinese book under the same title published by Higher Education Press (HEP), Beijing, China in 2007. We thank HEP for their kind collection of the errors found in the Chinese edition and suggestions from readers. In this English edition, some new materials have been written. For example, the necessary part of the uniqueness criterion is included.

After the first English draft was completed, the book was used as a textbook by Mathematics graduates. In the autumn of 2016, Professor Xian-Yuan Wu from Capital Normal University in Beijing invited Yong-Hua Mao to give a semester lecture on stochastic processes. It was a good chance to "test" the materials in the book.

Finally, we would thank Professor Zenghu Li for his suggestion to publish the book by World Scientific, to whom we also convey our thanks.

Mu-Fa Chen
Yong-Hua Mao
September 8, 2020

Preface to the Chinese Edition

Let us begin with a small probability event. When I traveled by TGV from Paris to London in November 2003, there were only three persons in our cabin: my wife, me, and a young Frenchman. Occasionally, I found that the young man seemed to be reading a Mathematics book, so I wondered he might also be studying Mathematics. I asked if he was reading a Mathematics book, and then looked at the cover of the book. It was a popular graduate text on Stochastic Processes (Stochastic Differential Equations). I was so glad thinking he studied probability theory. But he told me that he was working at a bank, not studying mathematics. Although having known that stochastic processes were very important in finance, I was greatly surprised at this situation. This is a small probability event, implying that stochastic processes have received its due attention and have many applications in science and engineering fields. I guessed that he may be "bullied" by the book, since he does not have the necessary foundation. I asked why he wanted to read such a difficult book. "I urge to know what is indeed stochastic integral and why it is useful for the finance", he answered. But I regretted my question, because I could not recommend a simpler book for him. On the other hand, I felt ashamed that as a probability researcher, I had not provided a common and simple book for those who were interested in it.

Although the current book is not a popular book, we hope to introduce the fundamentals of stochastic processes in a rather short length. For this, we selected the materials rather carefully. So we chose the two most important parts in stochastic processes—Markov chains and stochastic analysis. We wanted to lead the readers directly into the core of the topics, and more details are often left to a section called "Supplements and Exercises". So the book provides abundant exercises, more materials for further reading and studying. In the part of Markov chains, the core is the ergodicity. By using the minimal nonnegative solution method, the book deals with the recurrence and various ergodicity. This is done step by step, from finite state spaces to denumerable state spaces, from discrete time to continuous time. For the methods of proofs, we adopt the modern techniques, such as coupling technique and duality method. Some completely new results are included, such as the estimate of the spectral gap. The structure and proofs in the first part are rather different from the published textbooks on Markov chains. In the part of stochastic analysis, the book includes the martingale theory and Brownian motions, and the stochastic integral and stochastic differential equations emphasized in one dimension, while the multidimensional stochastic integral and stochastic equation are based on semimartingale. Three important topics—Feynman-Kac formula, random time transform and Girsanov's transform—are introduced. As an important application of the probability theory in classical mathematics, the book deals with the famous Brunn-Minkowski inequality in convex geometry. Usually probability theory is thought to be the study of randomness, however the modern probability theory is often used in different fields, such as MCMC, or even determinstic area: convex geometry and number theory. Some issues are involved in the book.

The textbook has taken a rather long time. For instance, the section "strong Markov property" followed the way I found in a lesson given in a probability seminar for teachers from the Chinese advanced normal colleges and universities. I gave an elective course for senior students twice in 1989 and 1996, and then in 1989, I gave a short course on Markov chains for the Chinese Mathematics Summer Schools and have written some lecture notes. Till 1999, I finished the first draft of this book. Afterwards, Professors Yong-Hua Mao and Ying-Zhe Wang have used it as a textbook for many years. In particular, Professor Mao made a lot of modifications, supplied Section 5 in Chapter 1, Section 4 in Chapter 2 and the whole of Chapter 7, and equipped all supplements and exercises for each chapter. In order not to spread out away from the stem, many details for the proofs and related contents are put into a section of supplements and exercises in each chapter. Like climbing a mountain, one can choose a winding path, a scenery a step, towards the peak, or one can take a "shortcut" to reach directly to the peak, to overlook the whole mountain and then enjoy the sceneries when going down the mountain.

We would like to thank many colleagues for using this book as their textbooks and for their numerous feedbacks and suggestions, especially colleagues from the School of Mathematical Science, Capital Normal University and the School of Mathematics and Computer Science, Anhui Normal University.

Mu-Fa Chen
December 5, 2006

Contents

PART 1
Markov Processes

Chapter 1

Discrete-Time Markov Chains

1.1 Stochastic Models for Economic Optimization and Markov Chains

We begin with an economic model.

Input-Output Method

To understand economic conditions, we start with the investigation. Firstly, we denote by $(x^{(1)}, x^{(2)}, \cdots, x^{(d)})$ the quantity of the main products we are interested in, which is called the **vector of products**. We fix the unit of the quantity of each product: kilogram, kilovolt and so on. We shall investigate three items. The first one is the initial vector of products (the input last year), denoted by

$$x_0 = (x_0^{(1)}, x_0^{(2)}, \cdots, x_0^{(d)}).$$

The second one is the output one year later:

$$x_1 = (x_1^{(1)}, x_1^{(2)}, \cdots, x_1^{(d)}).$$

The last one is the structure matrix (or matrix of expending coefficients) $A_1 = (a_{ij}^{(1)})$. This matrix is essential since it describes the efficiency of the current economy: to produce one unit of ith product, one needs $a_{ij}^{(1)}$ units of the jth product. That is, to produce $x_1^{(i)}$ units of the ith product, one needs $x_1^{(i)} a_{ij}^{(1)}$ units of the jth product. Thus the output x_1 expends

$$\sum_{i=1}^{d} x_1^{(i)} a_{ij}^{(0)}$$ units of the jth product. This is the input of the jth product

last year. Therefore $x_0^{(j)} = \sum_{i=1}^{d} x_1^{(i)} a_{ij}^{(1)}$. Write it in the vector form:

$$x_0 = x_1 A_1.$$

3

Matrix A_1 reflects the product efficiency this year, so it is called the **efficiency matrix, structure matrix,** or **expending coefficients.** Now, we consider the idealized model without consumption, however the model with consumption is closer to reality. But both models are mathematically similar, we will consider the latter model at the end of this section. In the idealized model, the vector of the products x_1 is wholly put into the reproduction and the output in the next year is x_2. In this case, as before we have $x_1 = x_2 A_2$. Inductively, we obtain

$$x_0 = x_n A_n \cdots A_1.$$

If we assume that $A_n \equiv A$ (the product efficiency is stable during a short period), then the above equation becomes $x_0 = x_n A^n$, $n \geqslant 0$. If A is invertible, then this implies

$$x_n = x_0 A^{-n}, \qquad n \geqslant 0. \tag{1.1}$$

In other words, once the initial vector of the products and efficiency matrix are known, we may predict the output x_n in the nth year. This is the famous **input-output method.** Up to 1960's, more than 100 countries have used this method in their national economy.

Method of Positive Eigenvector

The question is how to develop an economy well, or how fast the economy can grow? From (1.1), if the efficiency is stable (A fixed), then the question becomes: how to choose x_0 to make x_n grow as fast as possible? That is, we expect the ratios $x_1^{(j)}/x_0^{(j)}$ are as great as possible. This may be misleading, since for one j, this ratio is great enough, but for another j, this ratio may be small. To avoid this, we adopt the **minimax principle:** finding out the best solution among the worst cases. This is the safest strategy, and used widely in optimization theory and game theory. Namely, for a fixed efficiency matrix A, we find out x_0 such that

$$\max_{\substack{x_1 > 0 \\ x_0 = x_1 A}} \min_{1 \leqslant j \leqslant d} \frac{x_1^{(j)}}{x_0^{(j)}}$$

attains the maximum. Among other choices, the usual one is "average". If someone tells you that the average of the members' ages in a group is twenty, you may think that everyone in the group is strong, as in a team of volleyball players. However, the group may be a nursery, which consists of six babies and two older ladies, who are over seventy. The average of the ages in this group is still twenty. Another intrinsic reason for adapting the minimax principle is that, we expect each product increases according to a same ratio, to avoid the disproportion. This will become clear soon from the discussions below.

Now our question is: according to the above principle, how much is the optimal increasing rate? To answer this question, we need two concepts. A matrix or vector is called **nonnegative** or **positive**, if all the elements are nonnegative or positive. The above efficiency matrix A is nonnegative. A nonnegative matrix $A = (a_{ij})$ is called **irreducible** if for any $i \neq j$, there exist distinct $i_1, \cdots, i_m$ such that

$$a_{ii_1} a_{i_1 i_2} \cdots a_{i_{m-1} i_m} a_{i_m j} > 0.$$

Theorem 1.1 (L. K. Hua's fundamental theorem ([37])). Assume A is invertible, nonnegative and irreducible. Let u be the left eigenvector corresponding to its largest eigenvalue $\rho = \rho(A)$ (which is surely positive).

(1) If $x_0 = u$ (up to a positive factor), then $x_n = x_0 \rho^{-n}$, $n \geqslant 1$. In this case, it has the greatest increasing rate ρ^{-1}.

(2) If $0 < x_0 \neq u$ (up to a positive factor) and assume additionally A^{-1} is not nonnegative, then there exist n_0 and j_0 such that $x_{n_0}^{(j_0)} \leqslant 0$. In this case, the economy tends to collapse.

"Collapse" means that in some year, x_n contains a null or negative component, that is the **collapse time**

$$T := \inf\{n \geqslant 1 : \text{there exists some } j \text{ such that } x_n^{(j)} \leqslant 0\}$$

is finite. The above theorem gives the optimal production plan, or the best input plan. If T is always big enough, we do not care. However, let us look at the following example, due to L. K. Hua.

Example 1.2. Consider two products only: industry and agriculture. Let

$$A = \frac{1}{100} \begin{pmatrix} 25 & 14 \\ 40 & 12 \end{pmatrix}.$$

Then

$$u = \left(\frac{5}{7} (\sqrt{2409} + 13),\ 20 \right) \approx (44.34397483, 20).$$

For different inputs, the collapse times are listed in the following table.

x_0	T
$(44, 20)$	3
$(44.344, 20)$	8
$(44.34397483, 20)$	13

This shows that the economy is rather sensitive! The conclusion (2) is great! It shows that based on the current conditions, the economy has its own suitable development rate, neither too fast nor too slow. The reason behind is rather easy to understand. We can set up a lot of factories, but without enough electricity and communication, how can these factories operate? We can set up many universities, but without teachers and enough financial support, can they be universities with an empty library and no experiment equipments? We point out that this theorem is more essential than the Perron-Frobenius theorem or Brouwer fixed point theorem, which are often used in the study of economics. However they do not provide any information about the collapse phenomena.

Proof of L. K. Hua's Theorem

Of course, L. K. Hua's heuristic and important contribution is the second part. How to find out this assertion, it needs rather deep mathematical grounding, especially matrix theory. How is it related to **Markov chains**?

For Markov chains, the basic quantity is the **transition probability matrix** (or **stochastic matrix**) $P = (p_{ij})$. This matrix satisfies two conditions: all elements are nonnegative ($p_{ij} \geqslant 0$), and the sum of elements in each row equals 1 ($\sum_j p_{ij} = 1$). We have the following basic limit property.

Theorem 1.3. Let P be a (finite) positive stochastic matrix. Then limits

$$\lim_{n \to \infty} p_{ij}^{(n)} =: \pi_j > 0 \tag{1.2}$$

exist (independent of i) and $\sum_j \pi_j = 1$, where $P^n = (p_{ij}^{(n)})$ is the nth power of P.

Now, we prove Hua's theorem in the case of positive matrix by using Theorem 1.3.

Proof of Hua's Theorem (the case of positive matrix) a) Firstly, for the irreducible nonnegative matrix, we have the famous

Theorem 1.4 (Perron-Frobenius theorem). The greatest eigenvalue for the irreducible nonnegative matrix A is positive, and its corresponding left (row) vector u and right (column) eigenvector v are positive.

Without loss of generality, assume that $uv = 1$, that is $\sum_i u_i v_i = 1$.

b) Secondly, from $P^{n+1} = P^n P$, it follows that the distribution π in Theorem 1.3 satisfies equation $\pi = \pi P$, so that $\pi = \pi P^n$. From this and Theorem 1.3, π is the unique probability distribution satisfying equation

$\pi = \pi P$. Now, for the irreducible nonnegative matrix $A = (a_{ij})$, let

$$p_{ij} = \frac{a_{ij}v_j}{\rho(A)v_i}.$$

Then (p_{ij}) is a positive stochastic matrix. Because

$$\sum_i u_i v_i p_{ij} = \frac{1}{\rho(A)} \sum_i u_i a_{ij} v_j = u_j v_j$$

and $uv = 1$, applying Theorem 1.3, we know this stochastic matrix has the limit distribution $\pi_i = u_i v_i$. By induction, we have

$$p_{ij}^{(n)} = \frac{a_{ij}^{(n)}v_j}{\rho(A)^n v_i}.$$

Thus from Theorem 1.3 it follows that

$$\lim_{n \to \infty} \frac{a_{ij}^{(n)}}{\rho(A)^n} = \frac{v_i \pi_j}{v_j} = v_i u_j,$$

or in matrix form

$$\lim_{n \to \infty} \left(\frac{A}{\rho(A)} \right)^n = vu. \tag{1.3}$$

c) Thirdly, let $A > 0$. Then $A^{-1} \not\geq 0$. Without loss of generality, assume $\rho(A) = 1$ and $x_0 > 0$, $x_0 v = 1$. Then

$$Av = v \implies A^n v = v \implies v = A^{-n}v.$$

On the other hand, by $x_n = x_0 A^{-n}$ we have

$$x_n v = x_0 A^{-n} v = x_0 v = 1.$$

Thus if $x_n > 0$ holds for all n, then set $\{x_n : n \geq 1\}$ is bounded. Therefore its closure has limit points. Choose a subsequence $\{n_k\}$ such that

$$x_{n_k} \longrightarrow \text{some } \bar{x} \geq 0 \quad \text{and} \quad \bar{x}v = 1. \tag{1.4}$$

Finally, since

$$x_0 = \lim_{k \to \infty} x_0 A^{-n_k} A^{n_k} = \lim_{k \to \infty} x_{n_k} A^{n_k}$$

$$= \bar{x}vu \quad \text{(by (1.3) and (1.4))}$$

$$= u, \quad \text{(by (1.4) again)}$$

we see x_0 must be u ensuring $x_n > 0$ for every $n \geq 1$. $\square$

Notice in the proof, we only use (1.2). And condition "$A > 0$" comes from assumption "$P > 0$" in Theorem 1.3. In the next section, we will relax this condition. Now, let us go back to prove Theorem 1.3.

Proof of Theorem 1.3 Fix j and let

$$m_n = \min_i p_{ij}^{(n)}, \qquad M_n = \max_i p_{ij}^{(n)}.$$

The proof is completed by three steps:

1) $M_n \downarrow$;   2) $m_n \uparrow$;   3) $M_n - m_n \to 0$.

Prove 1) Let $\delta = \min_{i,j} p_{ij} > 0$. Choose $i_0 = i_0(n)$ such that $m_n = p_{i_0 j}^{(n)}$.
Then

$$M_{n+1} = \max_i \sum_k p_{ik} p_{kj}^{(n)} = \max_i \left[\sum_{k \neq i_0} p_{ik} p_{kj}^{(n)} + p_{ii_0} p_{i_0 j}^{(n)} \right]$$

$$\leqslant \max_i \left[\sum_{k \neq i_0} p_{ik} M_n + p_{ii_0} m_n \right] = \max_i \left[(1 - p_{ii_0}) M_n + p_{ii_0} m_n \right]$$

$$\leqslant M_n - (M_n - m_n)\delta.$$

The proof of 2) is similar:

$$m_{n+1} \geqslant m_n - (m_n - M_n)\delta.$$

Combining the previous two steps together, we get 3):

$$0 \leqslant M_n - m_n \leqslant (M_1 - m_1)(1 - 2\delta)^{n-1} \to 0. \quad \square$$

Indeed, the last step gives an estimate of the convergence rate for $p_{ij}^{(n)} - \pi_j$.

Stochastic Models

In practice, the efficiency matrix A is not deterministic, there always exists the randomness. For example, agriculture depends heavily on climate. Even ignoring the occasional factors, for big matrix A, its eigenvalue ρ is impossible to compute out exactly and the errors always exist. What do these factors affect? One may guess the randomness does not play a crucial role. Let us come back to the previous Example 1.2. In each year, replace $A = (a_{ij})$ by random matrix $A_n = \left(a_{ij}^{(n)} \right)$, where $a_{ij}^{(n)}$ is the same as the deterministic case a_{ij} with probability $2/3$. But $a_{ij}^{(n)}$ takes the value $(1 \pm 0.01)a_{ij}$ with probability $1/6$ respectively. Assume further all $a_{ij}^{(n)}$ $(i, j = 1, 2, n \geqslant 1)$ are independent. Starting from $x_0 = (44.344, 20)$, we obtain the probability of collapse time in each year:

n	1	2	3
probability	0	0.09	0.65

This shows the probability of collapse within 3 years is 0.74, which is rather different from the determining case $(T = 8)$. Thus, the production plan can be performed only for 2 years until it must be justified. The above discussions show that in the economic research, ignoring the randomness will cause distortion. This in some range explains why people think the input-output method is not good enough to apply. If the randomness is ignored, the distortion shall be great. Beginners studying probability theory often think probability theory gives a vague answer when people do not know the exact answer. Indeed, using stochastic mathematics will make the solution finer than deterministic ways.

In the randomness case, we have

Theorem 1.5 (Collapse theorem ([6])). Under suitable conditions, for every $x_0 > 0$, the collapse time is finite with probability 1, that is $\mathbb{P}^{x_0}[T = \infty] = 0$.

Here, we do not give the suitable conditions in the above theorem, however we would like to point out that the proof of the theorem is far from trivial. Its proof uses the limit theory of products of random matrices, which is now a very active field in the probability theory. In contract to the classical limit theory of sums of independent random variables, this new research topic is so attractive. For more details, see [32] and references therein. More recently, as an interaction field of the random matrix theory and operator algebra, the free probability has taken shape in the past decade, cf. [11; Chapter 10] for more information.

Remark 1.6. Professor L. K. Hua studied this problem first for the planned economy, however he pointed out that the model is also suitable for the market economy, by replacing the product quantity x_n by the market value quantity. More precisely, denote by p_i the market value of the ith product, and let V be the diagonal matrix with diagonal v_i/p_i (recall (v_i) is the right eigenvector corresponding to $\rho(A)$). Then we need just make the following transform

$$A \longrightarrow V^{-1}AV,$$
$$\rho(A) \longrightarrow \rho(V^{-1}AV) = \rho(A),$$
$$u \longrightarrow uV.$$

This shows the mathematical model remains unchanged. However, we should notice the randomness becomes greater in the market economy.

Remark 1.7. The model without consumption in this section is ideal, the more practical one is the model with consumption. Assume the increasing part in each product $x^{(i)}$ is taken out a proportion $\theta^{(i)} \in (0,1)$ for consumption. Then the vector of products one year later (the deterministic case) for reproduction is

$$y_1 = x_0 + (x_1 - x_0)(I - \Theta) = y_0[A^{-1}(I - \Theta) + \Theta], \qquad (1.5)$$

where $y_0 = x_0$, Θ is a diagonal matrix with diagonal $\{\theta^{(i)}\}$. Denote briefly $B = A^{-1}(I - \Theta) + \Theta$ and mimic the proof at the beginning of this section to derive the vector of product for reproduction in the nth year:

$$y_n = y_0 B^n, \qquad n \geqslant 0.$$

Mathematically, we just replace A^{-1} by B. However, the model with consumption is easier to be stable, which is consistent with the actual. ([15]).

Remark 1.8. The economic model with consumption should have its collapse theorem, but till now we only have a partial answer ([15]). On the other hand, the study for the collapse theorem is not the final aim, what we really expect is the optimal control theory. But the study in this direction is blank. It should be mentioned that the study of mathematical economy is far away from the practice. Economy is a much larger and complicated system, a huge field to be developed.

Remark 1.9. The method of positive eigenvector has become an important mathematical tool for the internet search engine. When searching in the internet, we fill in some key words. The search engine will search those related webpages, which are ranked and exported. The sort rule is from high rank to low one. Then how to define the ranks? The basis is the link between webpages (with suitable weights). Based on different methods, one can choose different link matrices between webpages, which are similar to the efficiency matrices in the economic models. The order of webpages is according to the positive eigenvector of this matrix. Obviously, the research on this topic has great value in application and commerce. Once we input the key words "search engine, page-rank" in

www.google.com,

we can find a lot of references. For example, we can get [45; 46].

Remark 1.10. We take the example of economic models to illustrate the applications of Markov chains. It is far away from being complete. In the economic models, because of the randomness in reality, we are forced to use stochastic mathematics. This is passive. However, the reverse situations appear, for some actual fields without randomness, people use their initiative to adopt stochastic mathematics. The typical example is the randomized algorithm, for which one of the most important tool is Markov chain. See [60] for example.

1.2   Discrete-Time Markov Chains. Recurrence and Ergodicity

We have seen from the previous section the importance of the limit of transition probability matrix " $\lim_{n \to \infty} p_{ij}^{(n)} = \pi_j$ ", and we have proved the limit in special case of "$P > 0$". The aim of this section is to deal with more general cases. We should point out: although this limiting property is purely analytical, the proofs for general cases need the probabilistic methods. In this section, we will see the power of the probabilistic intuition.

Markov Property

Assume that $E = \{i, j, k, \cdots\}$ is a finite or denumerable set.

Definition 1.11. Let $(X_n)_{n \geqslant 0}$ be a sequence of random variables defined on a probability space $(\Omega, \mathscr{F}, \mathbb{P})$, taking values in state space E. It is called a **Markov chain**, if the following Markov property holds: for any $i_0, i_1, \cdots, i_n$,

$$\mathbb{P}[X_n = i_n | X_0 = i_0, \cdots, X_{n-1} = i_{n-1}] = \mathbb{P}[X_n = i_n | X_{n-1} = i_{n-1}],$$

provided the conditional probability on the left-hand side is well-defined. And $p(n-1, i; n, j) := \mathbb{P}[X_n = j | X_{n-1} = i]$ is called **transition probability function** that the chain transits to state j at time n, starting from state i at time $n - 1$. If it is independent of n, called **homogeneous or time-homogeneous**, then denote it by p_{ij}.

The Markov property has the following equivalent statement (readers can prove it by themselves): Fix "present" $[X_m = i]$, then for every "past" $A \in \sigma(X_n : n < m)$ (the σ-algebra generated by $\{X_n : n < m\}$) and every "future" $B \in \sigma(X_n : n > m)$, it holds that

$$\mathbb{P}[AB | X_m = i] = \mathbb{P}[A | X_m = i]\, \mathbb{P}[B | X_m = i].$$

That is, conditional on "present", the "past" and "future" are independent. When deleting condition $[X_m = i]$, this becomes independence, rather than conditional independence. Therefore, "Markov property" is a correlation closest to "independence".

From now on, we restrict ourselves to homogeneous Markov chains. By Definition 1.11, we have a transition probability matrix:

$$P = (p_{ij}), \qquad p_{ij} \geqslant 0, \qquad \sum_{j \in E} p_{ij} = 1.$$

It is the one-step transition probability for the homogeneous Markov chain. Furthermore, by induction and Markov property, it is easy to see the nth power $P^n = \left(p_{ij}^{(n)}\right)$ of P is the n-step transition probability:

$$p_{ij}^{(n)} = \mathbb{P}[X_n = j | X_0 = i].$$

In particular, $P^0 = I$, that is, $p_{ij}^{(0)} = \delta_{ij}$. We call i leads to j, if there exist $i_1, \cdots, i_n$ such that $p_{ii_1} p_{i_1 i_2} \cdots p_{i_{n-1} i_n} p_{i_n j} > 0$, or, equivalently, there exists n such that $p_{ij}^{(n)} > 0$. If every state i can lead to every state $j \neq i$, then the probability matrix (or the Markov chain) is called **irreducible**.

Having the probabilistic interpretation for P, we can introduce various random variables or probabilistic quantities.

Recurrence and Positive Recurrence

Define the following two quantities.

Let us begin with the definition of **first return time**: $\tau_j^+ = \inf\{n \geq 1 : X_n = j\}$, $j \in E$. When the set is empty, define $\tau_j^+ = +\infty$. $[\tau_j^+ = \infty]$ means that the chain can never return to j. Here we use subscript "$+$", to differ from another random variable $\tau_j = \inf\{n \geq 0 : X_n = j\}$, which is called the **first hitting time**. From now on, we often abbreviate $\mathbb{E}[\cdot | X_0 = i]$ and $\mathbb{P}[\cdot | X_0 = i]$ to $\mathbb{E}_i$ and $\mathbb{P}_i$, respectively. Define

$$f_{ij}^{(n)} = \mathbb{P}_i[\tau_j^+ = n] = \mathbb{P}_i[X_n = j, X_m \neq j, 1 \leq m < n],$$

it denotes the probability that the chain returns to j exactly at the nth step starting from i. Then $f_{ij} := \sum_{n=1}^{\infty} f_{ij}^{(n)}$ is the probability that the chain returns to j after finitely many steps starting from i. Note that $\mathbb{P}_i[\tau_j^+ = \infty] = 1 - f_{ij}$.

Next define the **mean first return time** $\mathbb{E}_i \tau_j^+$:

$$m_{ij} = \mathbb{E}_i \tau_j^+ = \begin{cases} \sum_{n=1}^{\infty} n f_{ij}^{(n)}, & \text{if } f_{ij} = 1; \\ \infty, & \text{if } f_{ij} < 1. \end{cases}$$

Definition 1.12. A state j is called **recurrent** if $f_{jj} = 1$; otherwise it is called **transient**. State j is called **positive recurrent** if $\mathbb{E}_j \tau_j^+ < \infty$; if j is recurrent but not positive recurrent, then it is called **null recurrent**. Finally, we call **chain recurrent**, (**null recurrent**), or **positive recurrent**, if so are all states.

In view of applications, positive recurrence is much more important. Mathematically, $f_{ij}^{(n)}$ is most fundamental, since f_{ij} and $\mathbb{E}_i \tau_j^+$ can be expressed by it.

Roughly speaking, to obtain the nontrivial limit $\lim_{n\to\infty} p_{ij}^{(n)} > 0$, a necessary and sufficient condition is that P is (aperiodic and) positive recurrent. And for transience or null recurrence, $\lim_{n\to\infty} p_{ij}^{(n)} = 0$. This is just the probabilistic answer to the question we mentioned at the beginning of this section. To obtain these limits, we still have a long trip to go.

The following is a criterion for recurrence.

Theorem 1.13. A state i is recurrent if and only if (abbrev. iff) $\sum_{n=0}^{\infty} p_{ii}^{(n)} = \infty$.

When state i is transient, we have $\sum_{n=0}^{\infty} p_{ii}^{(n)} = (1 - f_{ii})^{-1}$.

The key for the proof is the following decomposition argument.

Lemma 1.14. For each i, j and $n \geqslant 1$, we have

$$p_{ij}^{(n)} = \sum_{m=1}^{n} f_{ij}^{(m)} p_{jj}^{(n-m)}. \tag{1.6}$$

Proof We will use the **method of first entrance**, that is, according to the first time reaching state j we make the decomposition (see the second equality below).

$$p_{ij}^{(n)} = \mathbb{P}_i[X_n = j] = \sum_{m=1}^{n} \mathbb{P}_i[\tau_j^+ = m, X_n = j]$$

$$= \sum_{m=1}^{n} \mathbb{P}_i[\tau_j^+ = m]\, \mathbb{P}_i[X_n = j | \tau_j^+ = m]$$

$$= \sum_{m=1}^{n} \mathbb{P}_i[\tau_j^+ = m]\, \mathbb{P}[X_n = j | X_0 = i, X_\nu \neq j, 1 \leqslant \nu < m, X_m = j]$$

$$= \sum_{m=1}^{n} f_{ij}^{(m)}\, \mathbb{P}[X_n = j | X_m = j] \quad \text{(by Markov property)}$$

$$= \sum_{m=1}^{n} f_{ij}^{(m)} p_{jj}^{(n-m)}, \qquad n \geqslant 1. \quad \square$$

Proof of Theorem 1.13 Note

$$\sum_{n=1}^{N} p_{ij}^{(n)} = \sum_{n=1}^{N} \sum_{m=1}^{n} f_{ij}^{(m)} p_{jj}^{(n-m)} \quad \text{(by (1.6))}$$

$$= \sum_{m=1}^{N} f_{ij}^{(m)} \sum_{n=m}^{N} p_{jj}^{(n-m)} = \sum_{m=1}^{N} f_{ij}^{(m)} \sum_{n=0}^{N-m} p_{jj}^{(n)}. \tag{1.7}$$

Take $j = i$ and let $N \to \infty$ to derive

$$\sum_{n=1}^{\infty} p_{ii}^{(n)} = f_{ii} \sum_{n=0}^{\infty} p_{ii}^{(n)} = f_{ii} \left(1 + \sum_{n=1}^{\infty} p_{ii}^{(n)}\right).$$

This implies the last assertion and the necessary part of the first assertion: $\sum_{n=0}^{\infty} p_{ii}^{(n)} < \infty \implies f_{ii} < 1$. However, the sufficient part of the first assertion $\left(\text{i.e. } \sum_{n=0}^{\infty} p_{ii}^{(n)} = \infty \implies f_{ii} = 1\right)$ can not be deduced directly from this equation, since it appears the problem of ∞/∞. To overcome the difficulty, we introduce the **generating functions**:

$$P_{ij}(s) = \sum_{n=0}^{\infty} s^n p_{ij}^{(n)}, \qquad F_{ij}(s) = \sum_{n=1}^{\infty} s^n f_{ij}^{(n)}, \qquad s \in (0,1).$$

(Note $f_{ij}^{(n)} \leqslant p_{ij}^{(n)} \leqslant 1$. These two function series converge on $s \in (0,1)$.) Then following the proof of (1.7), we deduce from (1.6) that

$$P_{ij}(s) - \delta_{ij} = P_{jj}(s)F_{ij}(s). \tag{1.8}$$

Thus $F_{ii}(s) = 1 - P_{ii}(s)^{-1}$. From this and

$$f_{ii} = \lim_{s \uparrow 1} F_{ii}(s), \qquad \sum_{n=0}^{\infty} p_{ii}^{(n)} = \lim_{s \uparrow 1} P_{ii}(s),$$

we obtain the assertion. $\square$

The above equation (1.8) reveals that the intrinsic relationship between limit $\lim_{n \to \infty} p_{ii}^{(n)}$ and probability quantity $\mathbb{E}_i[\tau_i^+]$:

$$\lim_{n \to \infty} p_{ii}^{(n)} = \lim_{s \uparrow 1}(1 - s)P_{ii}(s) \quad \text{(by Abel Theorem, easy)}$$

$$= \lim_{s \uparrow 1} \frac{1 - s}{1 - F_{ii}(s)} \quad \text{(by (1.8))}$$

$$= \lim_{s \uparrow 1} \frac{1}{F_{ii}'(s)} \quad \text{(if } i \text{ is recurrent)}$$

$$= 1/\mathbb{E}_i \tau_i^+.$$

The relationship is not obvious by intuition, but it is the starting point for the probabilistic analysis in this section.

We now discuss some simple properties for recurrence.

Theorem 1.15. For an irreducible Markov chain, all states are either recurrent or transient simultaneously.

Proof Take n and m such that $p_{ij}^{(n)} > 0$ and $p_{ji}^{(m)} > 0$. Then

$$p_{jj}^{(m+\nu+n)} \geqslant p_{ji}^{(m)} p_{ii}^{(\nu)} p_{ij}^{(n)}.$$

By Theorem 1.13, it is easy to see that if i is recurrent, then j must be recurrent. □

Corollary 1.16. Assume j is recurrent. If j leads to k, then $f_{kj} = 1$.

Proof Define

$$e_{jk}^{(n)} = \mathbb{P}_j[X_n = k, X_m \neq j, 0 < m < n]. \tag{1.9}$$

It denotes the probability that starting from j, the chain leads to k at the nth step without hitting j in the middle way. Since j can lead to k, there exists some n such that $e_{jk}^{(n)} > 0$. If $f_{kj} < 1$, then the probability that starting from j the chain never returns to j, is as follows

$$\mathbb{P}_j[\tau_j^+ = \infty] \geqslant e_{jk}^{(n)}(1 - f_{kj}) > 0.$$

This contradicts the recurrence of j. □

The following is our first limit theorem.

Theorem 1.17. Assume j is transient. Then for each $i \in E$, we have $\lim_{n\to\infty} p_{ij}^{(n)} = 0$.

Proof When $i = j$, the assertion comes from Theorem 1.13. When $i \neq j$, we use (1.6) and the dominated convergence theorem. The reason for this is that

$$p_{ij}^{(n)} = \sum_{m=1}^{n} f_{ij}^{(m)} p_{jj}^{(n-m)} = \sum_{m=1}^{\infty} f_{ij}^{(m)} p_{jj}^{(n-m)} I_{[m\leqslant n]}, \quad \sum_{m=1}^{\infty} f_{ij}^{(m)} = f_{ij} \leqslant 1,$$

and when $n \to \infty$

$$1 \geqslant p_{jj}^{(n-m)} I_{[m\leqslant n]} \to 0. \quad □$$

Periodicity

We have defined above the irreducibility. Note that for general state space E, it may contain several irreducible subsets. However, we can always restrict the chain in each of the irreducible subsets (still a Markov chain), and then deal with them one by one (the details are delayed in next section). Therefore we emphasize here on dealing with an irreducible class only.

Our aim is to study limit $\lim_{n\to\infty} p_{ij}^{(n)}$. But now, we may face a complicated situation. For example, consider a Markov chain on integers: $p_{i,i+1} > 0$, $p_{i,i-2} > 0$ and for others $p_{ij} = 0$ (one step forward, two steps backward!). Then starting from i, the chain can return to i in every three steps. That is, there exists a subsequence (n_k) such that $p_{ii}^{(n_k)} = 0$. To avoid this situation, we need the following concept.

Definition 1.18. Assume that set $\{n \geqslant 1 : p_{ii}^{(n)} > 0\}$ is not empty. Let d_i be the greatest common divisor of set $\{n \geqslant 1 : p_{ii}^{(n)} > 0\}$. d_i is called the **period** of state i. When $d_i = 1$, state i is called **aperiodic**. Furthermore, the chain is called aperiodic, if all states are aperiodic. An aperiodic positive recurrent chain is called **ergodic**.

Later we will prove the following property for aperiodicity (Proposition 1.30): State i is aperiodic if and only if there exists N such that $p_{ii}^{(n)} > 0$ whenever $n \geqslant N$.

In the matrix theory (in finite case), "irreducibility" is called "non-decomposition" (or "strong connection"); and "aperiodicity" is called "strong non-decomposition". The corresponding matrix is called "native matrix".

Limit Theorem

If we know that limit $\lim_{n\to\infty} p_{ij}^{(n)} = \pi_j$ exists, then by $P^{n+1} = P^n P$ and Fatou lemma, we have

$$\pi_j \geqslant \sum_i \pi_i p_{ij}. \qquad (1.10)$$

Thus

$$(1 \geqslant) \sum_j \pi_j \geqslant \sum_i \pi_i \sum_j p_{ij} = \sum_i \pi_i.$$

So the equality in (1.10) holds. Therefore we obtain equation $\pi_j = \sum_i \pi_i p_{ij}$, or $\pi = \pi P$. Furthermore,

$$\pi = \pi P^n, \qquad n \geqslant 0. \qquad (1.11)$$

For an irreducible chain, if $\pi_i > 0$ for some i, then $\pi_j > 0$ for every j. Indeed, since there exists n such that $p_{ij}^{(n)} > 0$, by (1.11) we see $\pi_j \geqslant \pi_i p_{ij}^{(n)} > 0$. Furthermore, by (1.11), we have

$$\pi_j = \sum_i \pi_i \lim_n p_{ij}^{(n)} = \left(\sum_i \pi_i\right) \pi_j, \qquad j \in E.$$

Thus we may assume $\sum_j \pi_j = 1$. Now, equation (1.11) shows that, if the initial distribution $\mathbb{P}[X_0 = i] = \pi_i$, then the distribution of the chain at each time n is the same: $\mathbb{P}[X_n = i] = \pi_i$. Based on these reasons, we call (π_i) the **stationary distribution** for the chain (X_n) or P. Another concept closely related to the stationary distribution is the invariant measure. $\mu = (\mu_i : i \in E)$ is called **invariant measure** for P, if $0 < \mu_i < \infty$ and $\mu = \mu P$.

The following is the core result for the discrete-time Markov chain, whose proof contains many probabilistic methods and techniques.

Theorem 1.19. Assume P is irreducible. The following statements are equivalent.

(1) P is positive recurrent.

(2) P has the stationary distribution π. Furthermore, the stationary distribution is unique: for every j, $\pi_j = m_{jj}^{-1}$, where $m_{ij} = \mathbb{E}_i \tau_j^+$.

If further P is aperiodic, then each of (1) and (2) is equivalent to

(3) $\lim\limits_{n \to \infty} p_{ij}^{(n)} = \pi_j > 0$.

This theorem tells us that in case of (aperiodic) positive recurrence, the limit of $p_{ij}^{(n)}$ is positive (this is why we call it positive recurrent and ergodic). On the other hand, Theorem 1.17 shows that the limit is null in a transient case. As for the case of null recurrence, we have

Theorem 1.20. Assume P is irreducible and null recurrent. Then for all i, $j \in E$, $\lim\limits_{n \to \infty} p_{ij}^{(n)} = 0$.

Proof of Theorem 1.19   Part A) Prove first (1) $\iff$ (2).

Set $e_{ji} = \sum\limits_{n=1}^{\infty} e_{ji}^{(n)}$, where $e_{ji}^{(n)}$ is defined by (1.9).

a) We will prove that if j is recurrent, then $\sum\limits_{i} e_{ji} = m_{jj} \, (= \mathbb{E}_j \tau_j^+])$.

For this, we need only use

$$\sum_i e_{ji}^{(n)} = \sum_i \mathbb{P}_j[X_n = i, X_m \neq j, 0 < m < n]$$

$$= \mathbb{P}_j[X_m \neq j, 0 < m < n] = \mathbb{P}_j[\tau_j^+ \geqslant n]$$

$$= 1 - f_{jj} + \sum_{m=n}^{\infty} f_{jj}^{(m)} \quad (\text{Note } \mathbb{P}_j[\tau_j^+ = \infty] = 1 - f_{jj} = 0)$$

$$= \sum_{m=n}^{\infty} f_{jj}^{(m)},$$

and then summation over n to derive the desired results.

b) Now prove that if j is recurrent, then for fixed j, $(\mu_i := e_{ji} : i \in E)$ is an invariant measure.

By definitions, we have

$$e_{ji}^{(1)} = p_{ji}, \qquad e_{ji}^{(n+1)} = \sum_{k \neq j} e_{jk}^{(n)} p_{ki}, \qquad n \geqslant 1.$$

Thus (μ_i) satisfies equation

$$\mu_i = \sum_{k \neq j} \mu_k p_{ki} + p_{ji} = \sum_k \mu_k p_{ki}. \tag{1.12}$$

In the last step we have used the recurrence: $\mu_j = e_{jj} = 1$. This gives us the invariant property and furthermore, we have $\mu = \mu P^n$ for every $n \geqslant 0$. By irreducibility, there exists n such that $p_{ij}^{(n)} > 0$. Thus

$$1 = \mu_j = \sum_k \mu_k p_{kj}^{(n)} \geqslant \mu_i p_{ij}^{(n)},$$

and hence $\mu_i \leqslant 1/p_{ij}^{(n)} < \infty$ for every $i \in E$.

c) Prove that for the irreducible recurrent chain, the invariant measure is unique, up to a factor.

Assume $\nu = \nu P$. Without loss of generality suppose (ν_i) is not null (that is, not all ν_i's are null). Then by $\nu = \nu P^n$ and irreducibility, every $\nu_i > 0$. Now we will adopt a dual technique. Namely, let $\tilde{p}_{ij} = \nu_j p_{ji}/\nu_i$. Then it is easy to prove that $\tilde{P} = (\tilde{p}_{ij})$ is a transition probability matrix (it is called a *dual* of P), and it is also irreducible. Furthermore, $\tilde{p}_{ij}^{(n)} = \nu_j p_{ji}^{(n)}/\nu_i$. Thus $\sum_{n=0}^{\infty} \tilde{p}_{ii}^{(n)} = \sum_{n=0}^{\infty} p_{ii}^{(n)} = \infty$, that is, $\tilde{P}$ is also recurrent. On the other hand, we make a decomposition according to "the last time reaching i before time n", to derive

$$p_{ij}^{(n)} = \sum_{m=0}^{n-1} \mathbb{P}_i[X_n = j, X_m = i, X_\ell \neq i, m < \ell < n]$$

$$= \sum_{m=0}^{n-1} \mathbb{P}_i[X_m = i]\, \mathbb{P}_i[X_n = j, X_\ell \neq i, m < \ell < n | X_m = i] \tag{1.13}$$

$$= \sum_{m=0}^{n-1} p_{ii}^{(m)} e_{ij}^{(n-m)}.$$

By using the generating functions in (1.13), we have

$$P_{ij}(s) - \delta_{ij} = P_{ii}(s)E_{ij}(s) \quad \left(E_{ij}(s) := \sum_{n=1}^{\infty} s^n e_{ij}^{(n)} \right).$$

Thus by b),

$$\lim_{s \uparrow 1} P_{ij}(s)/P_{ii}(s) = \lim_{s \uparrow 1} E_{ij}(s) = e_{ij} < \infty.$$

On the other hand, by (1.8) and Corollary 1.16, we have

$$\lim_{s \uparrow 1} P_{ij}(s)/P_{jj}(s) = f_{ij} = 1.$$

Apply it to $\tilde{P}$ to derive $\lim_{s\uparrow 1} \tilde{P}_{ij}(s)/\tilde{P}_{jj}(s) = 1$. Gather these two facts together and use the definition of $\tilde{P}$ to derive

$$1 = \lim_{s\uparrow 1} \tilde{P}_{ij}(s)/\tilde{P}_{jj}(s) = \frac{\nu_j}{\nu_i} \lim_{s\uparrow 1} P_{ji}(s)/P_{jj}(s) = \frac{\nu_j}{\nu_i} e_{ji}.$$

Thus $\nu_i/e_{ji} = \nu_j$, independent of i.

d) Finally prove that the positive recurrence $\iff$ the existence of stationary distribution.

Assume the chain is positive recurrent. Then by a) and b), we see $(\pi_i = e_{ji}/m_{jj} : i \in E)$ is the stationary distribution. In particular, $\pi_j = \mu_j/m_{jj} = m_{jj}^{-1}$. And, by c) this equality holds for every j. This proves (2). Conversely, assume that P has the stationary distribution π. Then by $\pi = \pi P^n$, Theorem 1.17 and the dominated convergence theorem, we know that the chain can not be transient. And then by b) and c), we see P has a unique invariant measure. Since $m_{jj} < \infty$ by a) and b), using again c), we have $\pi_j = m_{jj}^{-1}$ holds for all j.

Part B) Next prove (2) $\iff$ (3). In front of Theorem 1.19, we have proven (3) $\implies$ existence of the stationary distribution, so that we need only prove (2) $\implies$ (3).

a) Take two independent Markov chains $(X_n)_{n\geqslant 0}$ and $(Y_n)_{n\geqslant 0}$. They have the same transition probability matrix P, with different initial distributions. Fix i and let

$$\mathbb{P}[X_0 = i] = 1, \qquad \mathbb{P}[Y_0 = j] = \pi_j, \qquad j \in E.$$

Then by independence, $Z_n := (X_n, Y_n)$ has transition probability $P_{(i,j),(k,\ell)} = p_{ik} p_{j\ell}$. Furthermore, $p_{(i,j),(k,\ell)}^{(n)} = p_{ik}^{(n)} p_{j\ell}^{(n)}$ for every $n \geqslant 1$. Choose, respectively, n_{ik} and $n_{j\ell}$ such that $p_{ik}^{(n_{ik})} > 0$, $p_{j\ell}^{(n_{j\ell})} > 0$. By aperiodicity, we can choose $n_0 = n_0(k, \ell)$ such that $p_{kk}^{(n)} p_{\ell\ell}^{(n)} > 0$ for $n \geqslant n_0$. Thus, for $n \geqslant n_0$, we have

$$p_{(i,j),(k,\ell)}^{(n_{ik}+n_{j\ell}+n)} = p_{ik}^{(n_{ik}+n_{j\ell}+n)} p_{j\ell}^{(n_{ik}+n_{j\ell}+n)}$$
$$\geqslant p_{ik}^{(n_{ik})} p_{j\ell}^{(n_{j\ell})} p_{kk}^{(n+n_{j\ell})} p_{\ell\ell}^{(n+n_{ik})}$$
$$> 0.$$

So (Z_n) is also irreducible and aperiodic. It is easy to see (Z_n) has the stationary distribution $\pi_{(i,j)} = \pi_i \pi_j$. Let $\tau = \inf\{n \geqslant 0 : Z_n = (i,i)\}$. Then from d) of Part A), we know that existence of stationary distribution $\implies$ positive recurrence $\implies$ recurrence $\implies \mathbb{P}[\tau < \infty] = 1$. (In fact, it is enough to use the fact "irreducible stationary Markov chain is recurrent" proved in d) of Part A).)

b) On the other hand, for each j, we have

$$\mathbb{P}[X_n = j, \tau \leqslant n] = \sum_{m=0}^{n} \mathbb{P}[X_n = j, \tau = m] = \sum_{m=0}^{n} \sum_{k} \mathbb{P}[Z_n = (j,k), \tau = m]$$

$$= \sum_{m=0}^{n} \sum_{k} \mathbb{P}[\tau = m]\, \mathbb{P}[Z_n = (j,k)|\tau = m]$$

$$= \sum_{m=0}^{n} \sum_{k} \mathbb{P}[\tau = m]\, \mathbb{P}[Z_n = (j,k)|Z_\ell \neq (i,i), \ell < m, Z_m = (i,i)]$$

$$= \sum_{m=0}^{n} \sum_{k} \mathbb{P}[\tau = m]\, \mathbb{P}[Z_n = (j,k)|Z_m = (i,i)]$$

(by Markov property)

$$= \sum_{m=0}^{n} \sum_{k} \mathbb{P}[\tau = m]\, p_{ij}^{(n-m)} p_{ik}^{(n-m)}$$

$$= \sum_{m=0}^{n} \mathbb{P}[\tau = m]\, p_{ij}^{(n-m)}.$$

Similarly,

$$\mathbb{P}[Y_n = j, \tau \leqslant n] = \sum_{m=0}^{n} \mathbb{P}[\tau = m]\, p_{ij}^{(n-m)}.$$

Thus

$$|p_{ij}^{(n)} - \pi_j| = |\mathbb{P}[X_n = j] - \mathbb{P}[Y_n = j]|$$

$$= |\mathbb{P}[X_n = j, \tau > n] - \mathbb{P}[Y_n = j, \tau > n]|$$

$$\leqslant \mathbb{P}[\tau > n] \to 0, \qquad n \to \infty.$$

In the last step, we use the simple fact: $0 \leqslant a,\ b \leqslant c \Longrightarrow |a - b| \leqslant c.$ □

The proof of this theorem is to put two Markov chains into a same probability space. This is similar to the use of analytic geometry: choosing a suitable coordinate system for the study of different geometric objects. The bivariate process (Z_n) is called a **coupling** for (X_n) and (Y_n). This is the simplest coupling (**independent coupling**). Of course there are many important couplings rather than the independent coupling. The coupling method is one of the most important development of modern probability theory. For many attractive stories about the coupling method, refer to [10].

For an irreducible Markov chain, from proof of a)–c) in Part A), we know that if the chain has stationary distribution, then every state is positive recurrent. Conversely, if there is a positive recurrent state, then it follows from the above theorem that there exists stationary distribution, moreover every state is positive recurrent. Therefore we have the following assertion (similar to Corollary 1.15).

Corollary 1.21. For an irreducible recurrent Markov chain, all states are either positive recurrent or null recurrent simultaneously.

Proof of Theorem 1.20 a) Fix $i \neq j$. We still use the independent coupling in the proof of Theorem 1.19. Because there does not exist a stationary distribution π for (X_n), so does for (Z_n), and we change the initial condition for (Y_n): $\mathbb{P}[Y_0 = j] = 1$. If (Z_n) is transient, then by Theorem 1.17, we have

$$(p_{ij}^{(n)})^2 = p_{(i,i),(j,j)}^{(n)} \to 0.$$

This implies the desired assertion.

b) Now suppose (Z_n) is null recurrent. Following the second part of the proof of the previous theorem, we can obtain

$$\lim_{n\to\infty} \left| p_{ik}^{(n)} - p_{jk}^{(n)} \right| = 0, \qquad \forall k \in E. \tag{1.14}$$

Since for every $k \in E, \{p_{ik}^{(n)}\}_{n\geqslant 1}$ are bounded, there exists a convergence subsequence. Using the diagonal scheme method, we can choose a common subsequence $\{n_m\}$ such that $v_k = \lim_{m\to\infty} p_{ik}^{(n_m)}$ for every $k \in E$. And by (1.14), we have

$$v_k = \lim_{m\to\infty} p_{jk}^{(n_m)}, \quad k \in E.$$

Without loss of generality, assume (v_k) are not all null, otherwise the assertion is true. Obviously $\sum_k v_k \leqslant 1$. On the other hand, by the dominated convergence theorem and $P^{n_m+1} = PP^{n_m}$, we have

$$\lim_{m\to\infty} p_{jk}^{(n_m+1)} = v_k, \quad k \in E.$$

Then by $P^{n_m+1} = P^{n_m}P$ and Fatou lemma, we have

$$v_k \geqslant \sum_l v_\ell p_{lk}, \quad k \in E.$$

The argument used in front of Theorem 1.19 shows that the equality must hold. Thus P has a stationary distribution. By Theorem 1.19, the chain is positive recurrent, which is a contradiction. $\square$

It is easy to see that, an irreducible finite Markov chain is always recurrent. Furthermore, from the proof of Theorem 1.19 and $m_{jj} = \sum_i e_{ji}$, we know that the chain is indeed positive recurrent. In other words, a finite Markov chain can not be null recurrent. Note if P has period d, then P^d is aperiodic. Therefore the periodicity is not an essential difficulty. Thus we have proved L. K. Hua's theorem in Section 1 of this chapter, without assuming $P > 0$.

Generally speaking, the above definitions and criteria for recurrence and ergodicity are not easy to test. More efficient criteria are the theme of Section 4 in this chapter.

Remark 1.22. Theorem 1.13 has an alternative proof. Instead of the generating functions, we adopt approximation by finite sums. That is, we will use the following ratio limit theorem, which has its own interest.

Lemma 1.23. We have

$$\lim_{N \to \infty} \sum_{n=1}^{N} p_{ij}^{(n)} \bigg/ \sum_{n=0}^{N} p_{jj}^{(n)} = f_{ij}$$

for every i and j.

Alternative proof of Theorem 1.13 Indeed, by $p_{ij}^{(0)} = \delta_{ij}$ and the above equation

$$\lim_{N \to \infty} \left(1 - 1 \bigg/ \sum_{n=0}^{N} p_{ii}^{(n)} \right) = f_{ii}.$$

This implies Theorem 1.13.   □

Proof of Lemma 1.23 We need only apply the following Lemma 1.24 to

$$a_n = p_{jj}^{(n)}, \quad n \geqslant 0; \qquad b_n = \sum_{m=1}^{n} f_{ij}^{(m)}, \quad n \geqslant 1, \, b_0 = 0,$$

and use (1.7).   □

Lemma 1.24. Assume $(a_n : n \geqslant 0)$ is a nonnegative sequence, not all null, satisfying

$$\lim_{n \to \infty} a_n \bigg/ \sum_{m=0}^{n} a_m = 0.$$

Then

$$\lim_{n \to \infty} \sum_{m=0}^{n} a_m b_{n-m} \bigg/ \sum_{m=0}^{n} a_m = b$$

for every real sequence $(b_n : n \geqslant 0)$ such that $b := \lim_{n \to \infty} b_n$ exists (maybe $\pm\infty$).

Proof By assumption, for each $N \geqslant 1$ we have

$$\lim_{n \to \infty} \sum_{m=n-N+1}^{n} a_m \Big/ \sum_{m=0}^{n} a_m = 0. \tag{1.15}$$

a) Assume first $|b| < \infty$. For each $\varepsilon > 0$, take $N = N(\varepsilon)$ such that $|b_n - b| \leqslant \varepsilon$ whenever $n \geqslant N$. Thus there exists $B < \infty$ such that $|b_n - b| \leqslant B$ holds for all n. Furthermore,

$$\left| \sum_{m=0}^{n} a_m (b_{n-m} - b) \right| \leqslant \varepsilon \sum_{m=0}^{n-N} a_m + B \sum_{m=n-N+1}^{n} a_m.$$

Divided both sides by $\sum_{m=0}^{n} a_m$, from (1.15) it follows that

$$\varlimsup_{n \to \infty} \left| \sum_{m=0}^{n} a_m b_{n-m} \Big/ \sum_{m=0}^{n} a_m - b \right| \leqslant \varepsilon.$$

b) For $|b| = \infty$, without loss of generality assume $b = +\infty$. Then for each $M > 0$, there exists $N = N(M)$ such that when $n \geqslant N$, $b_n \geqslant M$. Thus

$$\sum_{m=0}^{n} a_m b_{n-m} \geqslant M \sum_{m=0}^{n-N} a_m + \Big(\min_{0 \leqslant n \leqslant N} b_n \Big) \sum_{m=n-N+1}^{n} a_m,$$

which implies the desired assertion.   □

Remark 1.25. Similarly, we can prove c) of the first part in Theorem 1.19 without the generating functions. Mimicking the proof of Lemma 1.23, by Lemma 1.24 and the proof b) in the first part of Theorem 1.19, we obtain

$$\lim_{N \to \infty} \sum_{n=0}^{N} p_{ij}^{(n)} \Big/ \sum_{n=0}^{N} p_{ii}^{(n)} = e_{ij} < \infty.$$

From this and Corollary 1.16 and Lemma 1.23 (apply it to $\tilde{P}$), it follows that

$$1 = \lim_{N \to \infty} \sum_{n=0}^{N} \tilde{p}_{ij}^{(n)} \Big/ \sum_{n=0}^{N} \tilde{p}_{jj}^{(n)} = \frac{\mu_j}{\mu_i} \lim_{N \to \infty} \sum_{n=0}^{N} p_{ji}^{(n)} \Big/ \sum_{n=0}^{N} p_{jj}^{(n)} = \frac{\mu_j}{\mu_i} e_{ji}.$$

Thus $\mu_i / e_{ji} = \mu_j$, independent of i.

Remark 1.26. Here, the proofs of Theorems 1.19 and 1.20 use the rather new dual method, so it has more probabilistic intuition. Their original proofs are much more analytic.

1.3   Limit Theorems in General Situation

For the particular class of irreducible and aperiodic Markov chains, we have seen the complete limit theorems in the last section. The purpose of this section is to extend these theorems to the general case. For this we need classify the state space in details.

Classification of States and Decomposition of State Space

Definition 1.27. Given $P = (p_{ij})$, we call i **leads directly to** j, if $p_{ij} > 0$; denote it by $i \to j$. We call i **leads to** j, if there exist $i_1, i_2, \cdots, i_m$ such that $i \to i_1 \to i_2 \to \cdots \to i_m \to j$; denote it by $i \rightsquigarrow j$. If $i \rightsquigarrow j$ and $j \rightsquigarrow i$, then we call i and j **communicate**. State i is called **essential**, if whenever $i \rightsquigarrow j$, then $j \rightsquigarrow i$; otherwise it is called **inessential**.

The communication relation is obviously transitive. Thus, according to this relation, we can classify the state space. We denote the equivalent class containing i by $C(i)$:

$$C(i) = \{i\} \cup \{j \neq i : i \text{ and } j \text{ communicate}\}.$$

The states in $C(i)$, if more than one, surely communicates with each other. However, it may happen that there exists $j \notin C(i)$ such that $i \rightsquigarrow j$ but $j \not\rightsquigarrow i$. That is, i may be inessential. If i is essential, then whenever $i \rightsquigarrow j$, j is also essential. Thus, $C(i)$ forms an irreducible **closed subset** of E:

$$\sum_{k \in C(i)} p_{jk} = 1, \qquad j \in C(i)$$

and it is the smallest closed subset, that is, there does not exist closed true subset of $C(i)$. By Lemma 1.16, we know every recurrent state is essential, thus $C(i)$ is closed for each recurrent state i. A special case is $p_{ii} = 1$, then by definitions, i is essential and $C(i) = \{i\}$. In this case, i is called an **absorbing state**.

If $\{n \geqslant 1 : p_{ii}^{(n)} > 0\}$ is empty, then state i is inessential. Thus, for the essential state, we can always define the **period** of a state i:

$$d_i = \text{the maximal divisor of set } \{n \geqslant 1 : p_{ii}^{(n)} > 0\}.$$

A basic result for the period is

Theorem 1.28. If i and j communicate, then $d_i = d_j$.

Thus, for an irreducible closed set, we can use a common period d. We have the following important property.

Proposition 1.29. Assume C is an irreducible closed set, with period $d > 1$. For every i and $j \in C$, if $p_{ij}^{(n_1)} > 0$ and $p_{ij}^{(n_2)} > 0$, then $n_1 - n_2$ can be divided by d.

According to these properties, the movement of the Markov chain in an irreducible closed set C is semi-deterministic with the periodicity rule. Namely, C can be decomposed into d disjoint sets:

$$C = \sum_{m=1}^{d} G_m.$$

Starting from a state in some G_m, the next step surely enters some state in $G_{m+1}(\mathrm{mod}\, d)$. In details, we first fix a state i, use Proposition 1.29 to make a decomposition according to the states the chain reaches step by step; then we can prove this decomposition is independent of the prefixed state i. Thus, if we view d steps as one step in the new chain, then this new chain becomes aperiodic. And in this new chain, each G_m now is closed. That is, the original chain can be decomposed into d irreducible and aperiodic new chains (sub-chains).

The above discussion shows that for an irreducible closed C with period d, every state i satisfies $p_{iC(i)}^{(nd)} := \sum_{j \in C(i)} p_{ij}^{(nd)} > 0$. This property can be partly generalized to

Proposition 1.30. Assume that the period of state i is d. Then there exists N such that $p_{ii}^{(nd)} > 0$ whenever $n \geqslant N$.

Summing up the above discussions, Corollaries 1.15 and 1.21, we obtain the following decomposition of the state space for a Markov chain, which also describes the full scene for movement of the system (Markov chain).

Decomposition of State Space and Sketch of Movement for the System

(1) If we regard an irreducible closed set as a class, then recurrence, null recurrence, positive recurrence and periodicity are all **class property**, that is, every state in a class has the same property.

(2) The union of null recurrent class $\{C_i\}$, ergodic classes $\{C_j\}$ and positive recurrent classes $\{C_k\}$, constitutes the recurrent set C. Starting from a state in some class C_s of C, the system will move inside this class C_s. In the periodic case, the system moves according to such a semi-determination rule.

(3) The other states constitute the transient set D. Starting from a state in D, the system either moves inside D forever, or enters and

stays inside some recurrent class.

Description of General Limit Theorem

Now, we can describe in details the limits for $p_{ij}^{(n)}$. Recall that $m_{ij} = \mathbb{E}_i \tau_j^+$; $\pi_j = m_{jj}^{-1}$ when j is recurrent.

Theorem 1.31.

   (1) If j is transient, then $\displaystyle \lim_{n \to \infty} p_{ij}^{(n)} = 0$ for each $i \in E$.

   (2) If j is recurrent with period $d = d_j$, then

$$\lim_{n \to \infty} p_{ij}^{(nd+r)} = \frac{d}{m_{jj}} \sum_{m=0}^{\infty} f_{ij}^{(md+r)} \tag{1.16}$$

for each i and $1 \leqslant r \leqslant d$. In particular, assume i is an essential state. If $i \notin C(j)$, then $p_{ij}^{(n)} = 0$ for every $n \geqslant 1$. If $i \in C(j)$, then $C(j) = C(i)$. Let j belong to rth subset G_r in the decomposition of $C(i)$. Then

$$p_{ij}^{(n)} = 0, \qquad \text{if } n \neq md + r; \tag{1.17}$$

$$\lim_{n \to \infty} p_{ij}^{(nd+r)} = \frac{d}{m_{jj}}. \tag{1.18}$$

This theorem looks complicated, the reason for this is the periodicity. If j is aperiodic, then this theorem becomes much simpler:

Corollary 1.32. If j is aperiodic, then

$$\lim_{n \to \infty} p_{ij}^{(n)} = f_{ij} \pi_j = f_{ij}/m_{jj} \tag{1.19}$$

for each $i \in E$. Note by convention, $m_{jj} = \infty$ (or $\pi_j = 0$) when j is transient.

If we replace the point-wise limit in n by the average limit, then the result becomes rather simpler.

Theorem 1.33. We have

$$\lim_{n \to \infty} \frac{1}{n} \sum_{m=1}^{n} p_{ij}^{(m)} = f_{ij} \pi_j \tag{1.20}$$

for every $i, j \in E$

Finally, notice for any stationary distribution π, $\pi = \pi P^n$ and moreover $\pi = \pi \left(\dfrac{1}{n} \displaystyle\sum_{m=1}^{n} P^m \right)$. By using Theorems 1.13, 1.19, 1.20 and 1.33, it is easy to prove (the reader may prove it by oneself) the following construction theorem for stationary distributions.

Theorem 1.34. Denote by $\{C_\alpha\}$ the total of positive recurrent irreducible closed sets. Let $H = E\backslash(\bigcup_\alpha C_\alpha)$. Then each stationary distribution for P takes the form:

$$\pi_i = \begin{cases} 0, & \text{if } i \in H; \\ \frac{\lambda_\alpha}{m_{ii}} = \lambda_\alpha \lim_{n\to\infty} \frac{1}{n} \sum_{m=1}^n p_{ii}^{(m)}, & \text{if } i \in C_\alpha, \end{cases}$$

where $\{\lambda_\alpha \geqslant 0\}$ satisfies $\sum_\alpha \lambda_\alpha = 1$.

Proofs

The rest of this section is devoted to prove the previous results. We will first prove the main theorems, Theorems 1.31 and 1.33, and then go back to prove the others. We will see that having at hand the basic theorems in the last section, the proofs for these theorems are not so difficult.

Proof of Theorem 1.31   a) Assertion (1) is just Theorem 1.13. From now on, assume that j is recurrent. In the special case that $i \notin C(j)$ and i is an essential state, it is clear that $p_{ij}^{(n)} = 0$ for every $n \geqslant 1$; otherwise, $i \rightsquigarrow j$ and then $j \rightsquigarrow i$ since i is essential. This leads to $i \in C(j)$, which contradicts the assumption $i \notin C(j)$.

b) Using the method of the first entrance, we have

$$p_{ij}^{(nd+r)} = \sum_{v=0}^{nd+r} f_{ij}^{(nd+r-v)} p_{jj}^{(v)} = \sum_{m=0}^n f_{ij}^{(md+r)} p_{jj}^{(n-m)d}, \qquad 1 \leqslant r \leqslant d. \tag{1.21}$$

In the last equality, we have used the periodicity of j: $p_{jj}^{(n)} = 0$ if $n \neq md$. On the other hand, applying Theorem 1.19 to the new chain with transition probability matrix $\tilde{P} = P^d$ on the irreducible closed set containing state j (it is aperiodic), we get $\lim_{n\to\infty} \tilde{p}_{jj}^{(n)} = 1/\tilde{m}_{jj}$. Since $\tilde{p}_{ij}^{(n)} = p_{ij}^{(nd)}$ and $\tilde{m}_{jj} = m_{jj}/d$, it follows that

$$\lim_{n\to\infty} p_{jj}^{(nd)} = d/m_{jj}. \tag{1.22}$$

Summing up (1.21) and (1.22), we arrive at (1.16).

c) If the essential state $i \in C(j)$ and j belongs to the rth subset in the decomposition of $C(i)$, then obviously $p_{ij}^{(n)} = 0$ holds for all $n \neq md + r$, which is (1.17). Of course, we then have $f_{ij}^{(n)} = 0$ holds for all $n \neq md + r$. Thus $\sum_{m=0}^\infty f_{ij}^{(md+r)} = \sum_{n=1}^\infty f_{ij}^{(n)} = f_{ij}$. Therefore, (1.18) follows from (1.16) and Corollary 1.16.   □

Proof of Theorem 1.33 When j is transient, the assertion follows from $\lim_{n\to\infty} p_{ij}^{(n)} = 0$ and $\pi_j = 0$. When j is recurrent, by (1.16),

$$\lim_{n\to\infty} p_{ij}^{(nd+r)} = d\pi_j \sum_{m=0}^{\infty} f_{ij}^{(md+r)}, \qquad 1 \leqslant r \leqslant d.$$

Thus, the desired assertion follows from the next Lemma 1.35 that

$$\lim_{n\to\infty} \frac{1}{n} \sum_{m=1}^{n} p_{ij}^{(m)} = \frac{1}{d} \sum_{r=1}^{d} d\pi_j \sum_{m=0}^{\infty} f_{ij}^{(md+r)} = f_{ij}\pi_j. \quad \square$$

Lemma 1.35. Suppose a postive integer d and a sequence $\{a_n : n \geqslant 1\}$ satisfy

$$\lim_{n\to\infty} a_{nd+r} = b_r, \qquad 1 \leqslant r \leqslant d.$$

Then

$$\lim_{n\to\infty} \frac{1}{n} \sum_{m=1}^{n} a_m = \frac{1}{d} \sum_{m=1}^{d} b_m.$$

Proof Since the average limit for a sequence coincides with the limit of the sequence, we have

$$\lim_{n\to\infty} \frac{1}{n} \sum_{m=1}^{n} a_{md+r} = b_r.$$

Set $A = \sup_{n\geqslant 1} a_n$, and denote $n = md + s$, $1 \leqslant s \leqslant d$. Then

$$\left| \frac{1}{n} \sum_{v=1}^{n} a_v - \frac{1}{d} \sum_{v=1}^{d} b_v \right| \leqslant \frac{1}{n} \sum_{v=md+1}^{md+s} a_v + \left| \frac{1}{n} \sum_{v=1}^{md} a_v - \frac{1}{d} \sum_{v=1}^{d} b_v \right|$$

$$\leqslant \frac{Ad}{n} + \frac{1}{d} \sum_{r=1}^{d} \left| \frac{md}{n} \frac{1}{m} \sum_{v=0}^{m-1} a_{vd+r} - b_r \right|.$$

Let $n \to \infty$, then $m \to \infty$ and $md/n \to 1$. This together with the previous equation gives the desired assertion. $\square$

Now, we go back to prove the results for periodicity at the beginning of this section.

Proof of Theorem 1.28 Assume $p_{ij}^{(m)} > 0$ and $p_{ji}^{(n)} > 0$ for some m, n. Then

$$p_{ii}^{(m+n)} \geqslant p_{ij}^{(m)} p_{ji}^{(n)} > 0,$$

so $d_i | (m + n)$. Similarly, $d_j | (m + n)$. If $p_{ii}^{(\ell)} > 0$, then

$$p_{jj}^{(m+n+\ell)} \geqslant p_{ji}^{(n)} p_{ii}^{(\ell)} p_{ij}^{(m)} > 0.$$

Thus $d_j | (m + n + \ell)$. These two facts show that $d_j | \ell$. Since this holds for every ℓ such that $p_{ii}^{(\ell)} > 0$ and in particular for d_i the maximal factor of all such ℓ, we have $d_j | d_i$. By the symmetry in i and j, we have $d_i | d_j$. Thus $d_i = d_j$. $\square$

Proof of Proposition 1.29   Since $j \rightsquigarrow i$, there exists m such that $p_{ji}^{(m)} > 0$. Then, by assumption, we have

$$p_{ii}^{(m+n_1)} \geqslant p_{ij}^{(n_1)} p_{ji}^{(m)} > 0, \qquad p_{ii}^{(m+n_2)} \geqslant p_{ij}^{(n_2)} p_{ji}^{(m)} > 0.$$

This shows $d_i | (m + n_1)$ and $d_i | (m + n_2)$. Therefore $d_i | (n_1 - n_2)$. $\square$

Proof of Proposition 1.30   The key is to use the following result in the element number theory: If positive integers $s_1, s_2, \cdots, s_m$ have the greatest common divisor d, then there exists positive integer N, such that for each $n \geqslant N$, there are nonnegative integers $c_1, c_2, \cdots, c_m$ such that $nd = \sum_{k=1}^{m} c_k s_k$.

Now choose $s_1, s_2, \cdots, s_m$, having the greatest common divisor d, such that $p_{ii}^{(s_k)} > 0$. By the above result, there exists N, such that for each $n \geqslant N$, there are nonnegative integers $\{c_k\}$ such that $nd = \sum_{k=1}^{m} c_k s_k$. Thus

$$p_{ii}^{(nd)} \geqslant \left(p_{ii}^{(s_1)} \right)^{c_1} \cdots \left(p_{ii}^{(s_m)} \right)^{c_m} > 0. \quad \square$$

Remark 1.36. Theorems 1.31 and 1.33 can be improved as follows. For a recurrent and irreducible closed set C with decomposition $\{G_r\}_{r=1}^d$, let

$$f(i, r, G_v) = \mathbb{P}_i [\text{there exists } n \equiv r \ (\text{mod } d) \text{ such that } X_n \in G_v],$$
$$f(i, C) = \mathbb{P}_i [\text{there exists } n \text{ such that } X_n \in C].$$

Then, (1.16) can be replaced by

$$\lim_{n \to \infty} p_{ij}^{(nd+r)} = \frac{d}{m_{jj}} f(i, r, G_v), \qquad \text{if } j \in G_v, \tag{1.23}$$

while (1.20) can be replaced by

$$\lim_{n \to \infty} \frac{1}{n} \sum_{m=1}^{n} p_{ij}^{(m)} = f(i, C) \pi_j, \qquad \text{if } j \in C. \tag{1.24}$$

The strength point of these results is that the coefficients rely on the class rather than the state. These conclusions can imply the following two results:

(1) $f(i, r, G_v) = \sum_{m=0}^{\infty} f_{ij}^{(md+r)}$ or each G_v and $1 \leqslant r \leqslant d$;

(2) $f(i, C) = f_{ij}$ or each $j \in C$.

Proof We only prove assertion (1), while the proof for assertion (2) is similar. Let $f_{ij}(r) = \sum_{m=0}^{\infty} f_{ij}^{(md+r)}$. Clearly, $f_{ij}(r) \leqslant f(i, r, G_v)$. Conversely, we have

$$f_{ij}(r) \geqslant \sum_{n=0}^{\infty} \sum_{k \in G_v} \mathbb{P}_i[X_{nd+r} = k,\ X_{vd+r} \neq k,\ 0 \leqslant v < n]\, f_{kj}(d).$$

Note that $f_{kj}(d) = f_{kj} = 1$ for $k, j \in G_v$. This implies the converse inequality. □

1.4   Criteria. The Minimal Nonnegative Solution

Criteria

As a continuation of the previous section, we present in this section some criteria, in terms of inequalities, to the topics studied in Section 1.2. The first one is for recurrence.

Theorem 1.37. Assume $P = (p_{ij})$ is irreducible. Let H be a non-empty finite subset of E. Then the chain is recurrent if and only if there exists finite nonnegative (y_i) such that

$$\begin{cases} \sum_{j \in E} p_{ij} y_j \leqslant y_i, & i \notin H, \\ \lim_{i \to \infty} y_i = \infty. \end{cases} \tag{1.25}$$

The last limit is taken by viewing E as $\mathbb{Z}_+ = \{0, 1, 2, \cdots\}$, but this is not loss of generality.

   Secondly, we use the same method to study ergodicity and convergence rates. The typical rate is of geometric convergence.

Theorem 1.38. Assume P is irreducible and aperiodic. Let H be a finite non-empty subset of E. Then the chain is positive recurrent (ergodic) iff inequalities

$$\begin{cases} \sum_{j} p_{ij} y_j \leqslant y_i - 1, & i \notin H, \\ \sum_{i \in H} \sum_{j} p_{ij} y_j < \infty \end{cases} \tag{1.26}$$

have a finite nonnegative solution.

   Typically, these criteria are applying to polynomials $y_i = c(1 + i^m)$.

   The chain is called **geometrically ergodic**, if there exists $\beta < 1$ such that for every i, j, n,

$$|p_{ij}^{(n)} - \pi_j| \leqslant c_{ij} \beta^n.$$

Theorem 1.39. Assume P and H are as above. Then the chain is geometrically ergodic iff there exists $\epsilon > 0$ such that inequalities

$$\begin{cases} \sum_j p_{ij} y_j \leqslant y_i(1 - \epsilon) - 1, & i \notin H, \\[2mm] \sum_{i \in H} \sum_j p_{ij} y_j < \infty \end{cases} \tag{1.27}$$

have a finite nonnegative solution.

The wonderful points for these theorems lie on the conditions that are determined mainly by $P = (p_{ij})$. A trick used in this criteria is the arbitrary finite subsets, which indicates that these properties depend essentially on the behavior of the chain at infinity. Their proofs should use the mathematical tool introduced below.

Theory of Minimal Nonnegative Solutions

In the study of Markov chains, we have to study infinitely many linear equations. Because they are not of finite dimension, the method of linear algebra is not always valid. We need to develop new methods. In 1970's, Chinese Mathematician Zhen-Ting Hou developed a mathematical tool, the theory of minimal nonnegative solutions. Refer to [33], and the materials below are adopted from [5].

We often meet with two kinds of equations

$$x_i = \sum_{k \in E} c_{ik} x_k + b_i, \qquad i \in E;$$

$$x_i(t) = \sum_{k \in E} \int_0^t c_{ik}(s) x_k(s) \mathrm{d}s + b_i(t), \qquad i \in E, \ t \geqslant 0,$$

where $c_{ik}, \ b_i \geqslant 0$.

In the following, we introduce an abstract form by using the notations below.

Let E be an arbitrary non-empty set and $\mathscr{H} \subset \{f : E \longrightarrow \bar{\mathbb{R}}_+ = [0, \infty]\}$ satisfy:

(1) $1 \in \mathscr{H}$;
(2) closed under the operation of nonnegative linear combination: $f_1, f_2 \in \mathscr{H}, \ c_1, c_2 \geqslant 0 \Longrightarrow c_1 f_1 + c_2 f_2 \in \mathscr{H}$;
(3) closed under the operation of increasing sequence limit: $f_n \in \mathscr{H}, \ f_n$ is increasing pointwise $\Longrightarrow \lim_n f_n \in \mathscr{H}$.

Operator $A : \mathscr{H} \longrightarrow \mathscr{H}$ satisfies conditions:
(1) $A0 = 0$;

(2) closed under the operation of nonnegative linear combination (cone mapping): $f_1, f_2 \in \mathscr{H}$, $c_1, c_2 \geqslant 0 \implies A(c_1 f_1 + c_2 f_2) = c_1 A f_1 + c_2 A f_2$;

(3) closed under the operation of increasing sequence limit: $f_n \in \mathscr{H}$, f_n is increasing pointwise $\implies A f_n \uparrow A \lim_n f_n$.

The total of operators which satisfy the above conditions is denoted by $\mathscr{A}$. For $A, \tilde{A}, A_n \in \mathscr{A}$, we call $A \leqslant \tilde{A}$ if $A f \leqslant \tilde{A} f$ for every $f \in \mathscr{H}$; and call $A_n \uparrow A$ if $A_n f \uparrow A f$ for every $f \in \mathscr{H}$.

Definition 1.40. Given $A \in \mathscr{A}$ and $g \in \mathscr{H}$, f^* is called the **minimal (nonnegative) solution** of equation

$$f = Af + g, \tag{1.28}$$

if f^* satisfies (1.28), and $\tilde{f} \geqslant f^*$ for every solution $\tilde{f}$ of (1.28).

Theorem 1.41 (Existence and uniqueness theorem). The minimal solution always exists and is unique. Namely, by letting

$$f^{(0)} = 0, \quad f^{(n+1)} = A f^{(n)} + g, \qquad n \geqslant 0, \tag{1.29}$$

we have $f^{(n)} \uparrow f^*$ as $n \to \infty$.

Proof Clearly, we have $f^{(n)} \in \mathscr{H}$ and $f^{(n)} \uparrow$ by induction. Then the limit exists (maybe $+\infty$). Thus, from

$$f^* = \lim_n f^{(n+1)} = \lim_n [A f^{(n)} + g] = A f^* + g,$$

we know that f^* is a solution. Now let $\tilde{f} \in \mathscr{H}$ be any solution of (1.28). Then $\tilde{f} \geqslant 0 = f^{(0)}$. Assume that $\tilde{f} \geqslant f^{(n)}$. Then

$$\tilde{f} = A\tilde{f} + g \geqslant A f^{(n)} + g = f^{(n+1)}.$$

Thus by induction, $\tilde{f} \geqslant f^{(n)}$ holds for every n, and hence $\tilde{f} \geqslant f^*$. This proves that f^* is minimal, obviously two minimal solutions must coincide.
□

We call equation (1.28) **homogeneous** if $g = 0$. In this case, it is easy to see that $f^* = 0$. The following examples are two simple equations and their minimal solutions.

$$x = x \implies x^* = 0; \qquad x = x + 2 \implies x^* = \infty.$$

Theorem 1.42 (Comparison theorem). Assume that $A, \tilde{A} \in \mathscr{A}$, and $g, \tilde{g} \in \mathscr{H}$ satisfy

$$\tilde{A} \geqslant A, \quad \tilde{g} \geqslant g.$$

Then for every $\tilde{f} \in \mathscr{H}$ satisfying $\tilde{f} \geqslant \tilde{A}\tilde{f} + \tilde{g}$, we have $\tilde{f} \geqslant f^*$ the minimal solution of (1.28).

Since the minimal solution is determined uniquely by A and g, we can denote it by $f^* = m_A(g)$.

Theorem 1.43 (Linear combination theorem). Assume G is countable and $\{a_s : s \in G\} \subset \overline{\mathbb{R}}_+$. Then

$$m_A\left(\sum_{s \in G} a_s g_s\right) = \sum_{s \in G} a_s m_A(g_s).$$

Theorem 1.44 (Monotone convergence theorem). Assume $\{A_n\} \subset \mathscr{A}$, $A_n \uparrow A$, $\{g_n\} \subset \mathscr{H}$, $g_n \uparrow g$. Then $A \in \mathscr{A}$, $g \in \mathscr{H}$ and $m_{A_n}(g_n) \uparrow m_A(g)$.

The iteration method used in Theorem 1.41 is called **the first iteration scheme**. In the following theorem, we have another iteration method, which is called **the second iteration scheme**.

Theorem 1.45. Assume $\{g_n\}_1^\infty \subset \mathscr{H}$. Let

$$\tilde{f}^{(1)} = g_1, \quad \tilde{f}^{(n+1)} = A\tilde{f}^{(n)} + g_{n+1}, \quad n \geqslant 1. \tag{1.30}$$

If $\sum_{k=1}^n g_k \uparrow g$ as $n \to \infty$, then $m_A(g) = \sum_{n=1}^\infty \tilde{f}^{(n)}$. In particular, if we let

$$\tilde{f}^{(1)} = g, \quad \tilde{f}^{(n+1)} = A\tilde{f}^{(n)}, \quad n \geqslant 1, \tag{1.31}$$

then

$$m_A(g) = f^* = \sum_{n=1}^\infty \tilde{f}^{(n)}.$$

The above results can be proved by using the induction method.

Lemma 1.46. Assume that c_{ij}, $b_i \in [0, \infty)$, i, $j \in E$.

(1) Let $(x_i^* : i \in E)$ be the minimal solution of equation

$$x_i = \sum_j c_{ij} x_j + b_i, \qquad i \in E.$$

Given $j \in E$, assume there exist $i = i_0, i_1, \cdots, i_n = j$ such that $c_{i_0 i_1} \cdots c_{i_{n-1} i_n} > 0$. We have

   i) $x_i^* < \infty \Longrightarrow x_j^* < \infty$;

   ii)   if further assume

$$\sum_j c_{ij} + b_i \leqslant 1, \quad i \in E, \tag{1.32}$$

then $x_i^* = 1 \Longrightarrow x_j^* = 1$.

In particular, if matrix (c_{ij}) is irreducible, then in case i), either $x_i^* = \infty$ for all $i \in E$, or $x_i^* < \infty$ for all i; in case ii), either $x_i^* = 1$ for all i, or $x_i^* < 1$ for all i.

(2) Fix $j_0 \in E$ and let $(x_i^* : i \in E)$ be the minimal solution of equation

$$x_i = \sum_{j \neq j_0} c_{ij} x_j + b_i, \qquad i \in E. \tag{1.33}$$

Then $x_i^* < \infty$ holds for all $i \in E$ if and only if $x_i^* < \infty$ holds for all $i \neq j_0$ and $\sum_{k \neq j_0} c_{j_0 k} x_k^* < \infty$. If

$$\sum_{j \neq j_0} c_{ij} + b_i = 1, \qquad i \in E,$$

then $x_i^* = 1$ holds for all $i \in E$ if and only if $x_i^* = 1$ holds for all $i \neq j_0$.

Proof a) Assume that $x_i^* < \infty$ and $c_{i i_1} c_{i_1 i_2} \cdots c_{i_{n-1} i_n} > 0$. Then by

$$\infty > x_i^* = \sum_k c_{ik} x_k^* + b_i \geqslant c_{i i_1} x_{i_1}^*,$$

we have $x_{i_1}^* < \infty$. By induction, we can prove that

$$x_{i_2}^* < \infty, \cdots, x_{i_n}^* < \infty$$

successively. Thus, the irreducibility implies that $x_j^* < \infty$ holds for all $j \in E$.

b) Suppose (1.32) holds. Clearly, by using the first iteration scheme, we have $x_i^* \leqslant 1$ for every $i \in E$. Let $x_i^* = 1$ and

$$c_{i i_1} c_{i_1 i_2} \cdots c_{i_{n-1} i_n} > 0.$$

If $x_{i_1}^* < 1$, then

$$1 = x_i^* = c_{i i_1} x_{i_1}^* + \sum_{j \neq i_1} c_{ij} x_j^* + b_i < c_{i i_1} + \sum_{j \neq i_1} c_{ij} x_j^* + b_i \leqslant \sum_j c_{ij} + b_i \leqslant 1.$$

This is a contradiction. Thus we have $x_{i_1}^* = 1$. By induction, we have

$$x_{i_2}^* = 1, \cdots, x_{i_n}^* = 1$$

successively. Hence the irreducibility implies that $x_i^* = 1$ holds for every $i \in E$.

c) If $(x_i^* : i \in E)$ is the minimal solution of equation (1.33), then it is easy to see that $(x_i^* : i \neq j_0)$ is the minimal solution of equation

$$x_i = \sum_{j \neq j_0} c_{ij} x_j + b_i, \quad i \neq j_0.$$

This implies the last assertion. □

The following are some element applications for the theory of minimal solution. More extensive and complicated applications are referred to [33].

Proposition 1.47. Given j, $(f_{ij} : i \in E)$ is the minimal solution of equation

$$x_i = \sum_{k \neq j} p_{ik} x_k + p_{ij}.$$

Proof We use the second iteration scheme in Theorem 1.45. Then

$$x_i^{(1)} = p_{ij} = f_{ij}^{(1)}.$$

By assuming $x_i^{(n)} = f_{ij}^{(n)}$, we have

$$x_i^{(n+1)} = \sum_{k \neq j} p_{ik} x_k^{(n)} = \sum_{k \neq j} p_{ik} f_{kj}^{(n)} = f_{ij}^{(n+1)}.$$

Hence $x_i^* = f_{ij}$. □

Proposition 1.48. Assume that the chain is recurrent. For fixed j, $(m_{ij} : i \in E)$ is the minimal solution of equation

$$x_i = \sum_{k \neq j} p_{ik} x_k + f_{ij}.$$

Proof Set

$$y_i^{(1)} = f_{ij}^{(1)}, \quad y_i^{(n+1)} = \sum_{k \neq j} p_{ik} y_k^{(n)} + f_{ij}^{(n+1)}, n \geq 1.$$

Assume that $y_i^{(n)} = n f_{ij}^{(n)} (n \geq 0)$. Then

$$y_i^{(n+1)} = n \sum_{k \neq j} p_{ik} f_{kj}^{(n)} + f_{ij}^{(n+1)} = n f_{ij}^{(n+1)} + f_{ij}^{(n+1)} = (n+1) f_{ij}^{(n+1)}.$$

Thus by Theorem 1.45 we have

$$x_i^* = \sum_{n-1}^{\infty} y_i^{(n)} = m_{ij}. \quad \square$$

In the following proof, we assume that H is a singleton, say $\{0\}$. For general finite sets H, the proofs are referred to [10; §4.3–§4.5].

Proof of Theorem 1.37 Define a new transition matrix $\tilde{P} = (\tilde{p}_{ij})$:

$$\tilde{p}_{ij} = \begin{cases} \delta_{ij} & i = 0; \\ p_{ij} & i \neq 0. \end{cases}$$

Hence 0 is an absorbing state, so that every $j \neq 0$ is not essential. Since a recurrent state is essential, it follows from Theorem 1.17 that $\lim_{n \to \infty} \tilde{p}_{ij}^{(n)} = 0$ for every $j \neq 0$. And it is easy to see that for $i \neq 0$, $\tilde{f}_{i0}^{(n)} = f_{i0}^{(n)}$ and $\tilde{p}_{i0}^{(n)} = \sum_{m=1}^{n} f_{i0}^{(m)}$, so that $\lim_{n \to \infty} \tilde{p}_{i0}^{(n)} = f_{i0}$.

a) *Sufficiency*   Let (y_i) be a solution to (1.25). Then

$$\sum_j \tilde{p}_{ij} y_j \leqslant y_i, \quad i \in E,$$

and for $n \geqslant 1$

$$\sum_j \tilde{p}_{ij}^{(n)} y_j \leqslant y_i, \quad i \in E.$$

Thus

$$1 = \sum_{j \leqslant N} \tilde{p}_{ij}^{(n)} + \sum_{j > N} \tilde{p}_{ij}^{(n)} \leqslant \sum_{j \leqslant N} \tilde{p}_{ij}^{(n)} + \sum_{j > N} \tilde{p}_{ij}^{(n)} \frac{y_j}{\inf_{k > N} y_k}$$

$$\leqslant \sum_{j \leqslant N} \tilde{p}_{ij}^{(n)} + \frac{y_i}{\inf_{k > N} y_k}.$$

By letting firstly $n \to \infty$, and then $N \to \infty$, we have $f_{i0} = 1$ for every $i \neq 0$. Apply Proposition 1.47 to P, we obtain $f_{00} = \sum_{j \neq 0} p_{0j} f_{j0} + p_{00} = 1$.
Therefore P is recurrent.

b) *Necessity*   Denote by $\tilde{X}_n$ the chain corresponding to $\tilde{P}$, and define

$$\tilde{\eta}_i(n) = \mathbb{P}_i[\exists\, m \geqslant 0 \text{ such that } \tilde{X}_m \geqslant n].$$

Then $\tilde{\eta}_0(n) = 0$ for each $n > 0$, and $\tilde{\eta}_i(n) = 1$ for each $i \geqslant n$. Since P is recurrent and irreducible, $f_{i0} = 1$ for every $i \geqslant 0$. Hence

$$\lim_{n \to \infty} \tilde{\eta}_i(n) = 0, \quad i \geqslant 0.$$

Choose $n_k \uparrow$ such that $\tilde{\eta}_i(n_k) < 2^{-k}$, $i \leqslant k$. Define $y_i = \sum_{k \geqslant 1} \tilde{\eta}_i(n_k) < \infty$.
Then by Fatou's lemma

$$\varliminf_{i \to \infty} y_i = \varliminf_{i \to \infty} \sum_{k \geqslant 1} \tilde{\eta}_i(n_k) \geqslant \sum_{k \geqslant 1} \varliminf_{i \to \infty} \tilde{\eta}_i(n_k) = \infty.$$

By using Markov property, it is easy to check that

$$\tilde{\eta}_i(n) = \mathbb{P}_i[\exists m \geqslant 0 \text{ such that } \tilde{X}_m \geqslant n]$$

$$\geqslant \sum_j \mathbb{P}_i[\exists m \geqslant 1 \text{ such that } \tilde{X}_m \geqslant n, \tilde{X}_1 = j]$$

$$= \sum_j \mathbb{P}_i[\tilde{X}_1 = j]\, \mathbb{P}_j[\exists m \geqslant 0 \text{ such that } \tilde{X}_m \geqslant n]$$

$$= \sum_j \tilde{p}_{ij} \tilde{\eta}_j(n).$$

Thus for $i \neq 0$,

$$y_i = \sum_{k \geqslant 1} \tilde{\eta}_i(n_k) \geqslant \sum_j \tilde{p}_{ij} y_j = \sum_j p_{ij} y_j. \quad \square$$

Theorem 1.49. Assume that P is irreducible. Then the chain is transient if and only if equation

$$\sum_{j \in E} p_{ij} y_j = y_i, \quad i \notin 0$$

has a non-constant bounded solution.

Proof Consider Markov chain $\tilde{P}$ defined in the proof of Theorem 1.37. Suppose that the chain is transient, then there exists $i \neq 0$ such that $\tilde{f}_{i0} = f_{i0} < 1$. Since $\tilde{f}_{00} = 1$, it is easy to check

$$\sum_{j \in E} \tilde{p}_{ij} \tilde{f}_{j0} = \tilde{f}_{i0}, \quad i \in E.$$

That is, $(y_i = \tilde{f}_{i0} : i \in E)$ is the desired non-constant bounded solution.

Conversely, suppose the equation has a bounded solution (y_i), namely $y_i \leqslant M < \infty$. Note that

$$\sum_{j \in E} \tilde{p}_{ij} y_j = y_i, \quad i \in E$$

and then

$$\sum_{j \in E} \tilde{p}_{ij}^{(n)} y_j = y_i, \quad i \in E, n \geqslant 1.$$

If the chain is recurrent, then $\lim_{n \to \infty} \tilde{p}_{i0}^{(n)} = f_{i0} = 1$. Hence

$$\sum_{j \neq 0} \tilde{p}_{ij}^{(n)} y_j \leqslant M \sum_{j \neq 0} \tilde{p}_{ij}^{(n)} \leqslant M(1 - \tilde{p}_{i0}^{(n)}) \to 0 \quad \text{as} \quad n \to \infty.$$

But

$$y_i = \sum_{i \neq 0} \tilde{p}_{ij}^{(n)} y_j + \tilde{p}_{i0}^{(n)} y_0 \to y_0 \quad \text{as} \quad n \to \infty.$$

This shows that (y_i) must be constant. $\square$

Proof of Theorem 1.38 Now suppose the chain is positive recurrent. Then it follows from Proposition 1.48 that

$$m_{i0} = \sum_{k \neq 0} p_{ik} m_{k0} + 1, \quad i \in E,$$

or

$$\sum_{k \neq 0} p_{ik} m_{k0} = m_{i0} - 1, \quad i \in E. \tag{1.34}$$

Take $y_0 = 0$ and $y_i = m_{i0}$ for $i \neq 0$. Then when $i \neq 0$, the conditions for Theorem 1.38 are fulfilled. And when $i = 0$, it follows from positive recurrence that $m_{00} < \infty$. Hence by (1.34) we have

$$\sum_k p_{0k} y_k = \sum_{k \neq 0} p_{0k} m_{k0} = m_{00} - 1 < \infty.$$

Thus all conditions in the theorem are fulfilled. This proves the necessity in Theorem 1.38.

Conversely, suppose (y_i) is a solution to inequalities (1.26). Let $c = \sum_j p_{0j} y_j < \infty$ and define

$$y_i^{(1)} = y_i \text{ and } y_i^{(n+1)} = \sum_j p_{ij}^{(n)} y_j \text{ for } n \geqslant 1, \ i \in E.$$

Then $y_0^{(2)} = c$ and $y_j^{(2)} \leqslant y_j - 1$ for $j \neq 0$. Furthermore,

$$
\begin{aligned}
y_i^{(n+2)} &= \sum_j p_{ij}^{(n+1)} y_j = \sum_{j,k} p_{ik}^{(n)} p_{kj} y_j \\
&= \sum_j p_{ij}^{(n)} y_j^{(2)} \leqslant c p_{i0}^{(n)} + \sum_{j \neq 0} p_{ij}^{(n)} (y_j - 1) \\
&= c p_{i0}^{(n)} + \sum_j p_{ij}^{(n)} y_j - \sum_{j \neq 0} p_{ij}^{(n)} - p_{i0}^{(n)} y_0 \\
&\leqslant (1 + c) p_{i0}^{(n)} + y_i^{(n+1)} - 1.
\end{aligned}
$$

By induction, we have $y_i^{(n)} < \infty$ for all n and i. By induction again, we obtain

$$y_i^{(n+2)} \leqslant (1 + c) \sum_{r=1}^n p_{i0}^{(r)} + y_i^{(2)} - n,$$

or

$$\frac{y_i^{(n+2)}}{n} \leqslant (1 + c) \frac{1}{n} \sum_{r=1}^n p_{i0}^{(r)} + \frac{y_i^{(2)}}{n} - 1.$$

By letting $n \to \infty$, from Theorem 1.33 it follows that

$$
\begin{aligned}
&0 \leqslant (1 + c) f_{i0} \pi_0 - 1 \\
&\Longrightarrow f_{i0} \pi_0 \geqslant (1 + c)^{-1} > 0 \\
&\Longrightarrow \pi_0 > 0 \\
&\Longrightarrow \pi_j > 0, \quad j \in E \qquad \text{(by irreducibility).} \quad \square
\end{aligned}
$$

1.5   Some Typical Discrete-Time Markov Chains

In this section, we study the recurrence, ergodicity for some typical Markov chains: random walks, branching processes and models in queueing theory. Each of them has strong background.

Random Walks

The following Markov chain P on $E = \mathbb{Z}_+$, is called a **random walk**: $p_{i,i+1} = p_i$, $p_{i+1,i} = q_{i+1}$, $p_{ii} = r_i (i \geqslant 0), p_i + q_i + r_i = 1 (i \geqslant 1)$ and $p_0 + r_0 = 1$.

Assume $p_i > 0$, $r_i \geqslant 0$ $(i \geqslant 0)$ and $q_i > 0$ $(i \geqslant 1)$. Obviously, if $r_i = 0$ for every $i \geqslant 0$, the chain has period 2; otherwise, then it is aperiodic. Assume that $r_0 > 0$, then the chain is aperiodic. The chain is also called a random walk on the half line. Define

$$\mu_0 = 1, \quad \mu_i = \frac{p_0 p_1 \cdots p_{i-1}}{q_1 q_2 \cdots q_i}, \quad i \geqslant 1. \tag{1.35}$$

Next we apply the results obtained in the previous sections to give some criteria for recurrence and ergodicity of random walks.

Theorem 1.50.   (1) The chain is recurrent iff $\sum\limits_{i=0}^{\infty} (\mu_i p_i)^{-1} = \infty$;

(2) The chain is ergodic iff

$$\mu := \sum_{i=0}^{\infty} \mu_i < \infty.$$

And in case (2), the stationary distribution is $\pi_i = \mu_i/\mu$ $(i \geqslant 0)$.

Proof To prove (1), consider equation

$$\sum_{j \geqslant 0} p_{ij} y_j = y_i, \quad i \neq 0.$$

That is,

$$q_1 y_0 + r_1 y_1 + p_1 y_2 = y_1,$$

$$q_2 y_1 + r_2 y_2 + p_2 y_3 = y_2, \tag{1.36}$$

$$\cdots\cdots\cdots$$

By applying Theorem 1.49, to prove assertion (1), we need only to prove equation (1.36) has non-constant bounded solution iff $\sum\limits_{i=0}^{\infty} (\mu_i p_i)^{-1} < \infty$. For this, we prove

$$y_0 = 0, \quad y_n = \sum_{i=0}^{n-1} 1/\mu_i p_i, \quad i \geqslant 1$$

is the non-constant bounded solution of equation (1.36). Indeed, since $r_n = 1 - p_n - q_n$, it follows that

$$q_n y_{n-1} + r_n y_n + p_n y_{n+1} = q_n \sum_{i=0}^{n-2} \frac{1}{\mu_i p_i} + r_n \sum_{i=0}^{n-1} \frac{1}{\mu_i p_i} + p_n \sum_{i=0}^{n} \frac{1}{\mu_i p_i}$$

$$= -\frac{q_n}{\mu_{n-1} p_{n-1}} + \frac{p_n}{\mu_n p_n} + \sum_{i=0}^{n-1} \frac{1}{\mu_i p_i} = y_n.$$

Now we prove (2). It is easy to check that $(\mu_i : i \geqslant 0)$ satisfies $\mu_i p_{ij} = \mu_j p_{ji}$, so that $\sum_{i \geqslant 0} \mu_i p_{ij} = \mu_j$. Thus, by definition, $\pi_i = \mu_i / \mu$ is the stationary distribution of P once $\mu < \infty$. Hence the random walk is positive recurrent iff $\mu < \infty$. $\square$

Queueing Theory

The transition probability matrix $P = (p_{ij})$ of discrete-time queueing model is as follows:

$$p_{ij} = \begin{cases} a_j, & \text{if } i = 0; \\ a_{j-i+1}, & \text{if } j \geqslant i - 1; \\ 0, & \text{else,} \end{cases}$$

where $a_k > 0$, $\sum_{k=0}^{\infty} a_k = 1$. Let $\rho = \sum_{k=0}^{\infty} k a_k$ and $f(z) = \sum_{k=0}^{\infty} a_k z^k$ for $z \in [0, 1]$. We are going to prove that

Theorem 1.51. The chain is recurrent iff $\rho \leqslant 1$; while it is positive recurrent (ergodic) iff $\rho < 1$.

Proof a) If $\rho > 1$, then we prove equation $\sum_{j \geqslant 0} p_{ij} y_j = y_i$, $i \neq 0$ has the non-constant bounded solution. Let $y_i = c^i$. Then from this equation we have

$$\sum_{j \geqslant 0} p_{ij} c^j = \sum_{j \geqslant i-1} a_{j-i+1} c^j = c^i,$$

that is,

$$c = \sum_{j \geqslant i-1} a_{j-i+1} c^{j-i+1} = \sum_{k \geqslant 0} a_k c^k =: f(c).$$

Since

$$f(0) < 1 = f(1) \text{ and } f'(1) = \rho > 1,$$

by the convexity and monotonicity of f on $[0,1]$, there exists unique $c_0 \in (0,1)$ such that $f(c_0) = c_0$. Thus $y_i = c_0^i$ is the non-constant bounded solution to the equation.

b) If $\rho \leqslant 1$, then by taking $y_j = j$,

$$\sum_{j \geqslant 0} j p_{ij} = \sum_{j \geqslant i-1} j a_{j-i+1} = \sum_{j \geqslant i-1} (j-i+1) a_{j-i+1} + i - 1$$

$$= \rho - 1 + i \leqslant i$$

for $i \neq 0$. From Theorem 1.37 it follows that the chain is recurrent.

c) If $\rho < 1$, then by taking $y_i = i/(1-\rho)$, as proved in b) we have

$$\sum_{j \geqslant 0} p_{ij} y_j = y_i - 1, \quad i \neq 0.$$

Thus from Theorem 1.38 it follows that the chain is positive recurrent.

d) Suppose the chain is ergodic, namely, there is the stationary distribution $(\pi_i : i \geqslant 0)$. Let $\Pi(z) = \sum_{j \geqslant 0} \pi_j z^j$ for $z \in (0,1)$. Since

$$\pi_j = \sum_i \pi_i p_{ij} = \pi_0 a_j + \sum_{i=1}^{j+1} \pi_i a_{j-i+1},$$

we have

$$\Pi(z) = \sum_{j \geqslant 0} \pi_0 a_j z^j + \sum_{j \geqslant 0} z^j \sum_{i=1}^{j+1} \pi_i a_{j-i+1} = \pi_0 f(z) + \frac{1}{z}(\Pi(z) - \pi_0)f(z).$$

Thus,

$$\frac{f(z) - 1}{z - 1} = 1 - \pi_0 \frac{f(z)}{\Pi(z)}.$$

Let $z \uparrow 1$ to derive that $\rho = f'(1) = 1 - \pi_0 < 1$.

e) Furthermore, we actually obtain an expression for the stationary distribution, that is,

$$\Pi(z) = \frac{(1-\rho)(1-z)f(z)}{f(z) - z}, \quad z \in (0,1). \quad \square$$

Branching Processes

Suppose $\{Y, Y_{n,k} : n, k \geqslant 1\}$ is a sequence of independent and identically distributed random variables taking values on $\mathbb{Z}_+$, with $\mathbb{P}[Y = i] = a_i$ ($i = 0, 1, 2, \cdots$). Let

$$X_{n+1} = \sum_{k=1}^{X_n} Y_{n+1,k}, \quad n \geqslant 0.$$

Then $(X_n)_{n\geqslant 0}$ is a Markov chain, with transition probability matrix $P = (p_{ij})$:

$$p_{ij} = \frac{1}{j!}\frac{\partial^j}{\partial x^j}\left[\left(\sum_{k=0}^{\infty}a_k x^k\right)^i\right]\Bigg|_{x=0}.$$

Obviously $p_{00} = 1$, that is, 0 is an absorbing state. The problem is the **extinction probability** $f_{i0} =$? The main result for the branching processes is as follows.

Theorem 1.52.

(1) $f_{i0} = (f_{10})^i$ for $i \geqslant 0$;

(2) Let $G(z) = \sum_{i=0}^{\infty} a_i z^i$, $z \in [0,1)$. Then f_{10} is the smallest positive root of equation $z = G(z)$. And $f_{10} = 1$ when $\rho := \sum_{k=0}^{\infty} ka_k \leqslant 1$; $f_{10} \in (0,1)$ when $\rho > 1$.

Proof Note that function $G(z)$ is the generating function of random variable Y. It is easy to prove that if random variables ξ, η are independent, then their generating functions G_ξ, G_η and $G_{\xi+\eta}$ have relation: $G_{\xi+\eta}(z) = G_\xi(z)G_\eta(z)$. Let $X_0 = i$. Then we have

$$G_{i,n+1}(z) := \sum_{k=0}^{\infty} z^k \mathbb{P}_i[X_{n+1} = k]$$

$$= \sum_{k=0}^{\infty} z^k \sum_{j=0}^{\infty} \mathbb{P}_i[X_{n+1} = k | X_n = j]\,\mathbb{P}_i[X_n = j]$$

$$= \sum_{k=0}^{\infty} z^k \sum_{j=0}^{\infty} \mathbb{P}\left[\sum_{m=1}^{j} Y_{n+1,m} = k\right] \mathbb{P}_i[X_n = j]$$

$$= \sum_{j=0}^{\infty} \mathbb{P}_i[X_n = j] \sum_{k=0}^{\infty} z^k \mathbb{P}\left[\sum_{m=1}^{j} Y_{n+1,m} = k\right]$$

$$= \sum_{j=0}^{\infty} \mathbb{P}_i[X_n = j]G(z)^j = G_{i,n}(G(z)), \quad n \geqslant 1,$$

and $G_{i,1}(z) = G(z)^i$. Thus by induction we have $G_{1,n+1}(z) = G(G_{1,n}(z))$ and

$$G_{1,n}(z) = \underbrace{G(G(\cdots(G(z))\cdots))}_{n \text{ folds}}, \quad G_{i,n}(z) = G_{1,n}(z)^i.$$

Notice $p_{i0}^{(n)} = G_{i,n}(0)$ and $f_{i0} = \lim_{n \to \infty} p_{i0}^{(n)}$. Then we have $f_{i0} = (f_{10})^i$. We can and do assume that $X_0 = 1$. Set $z_0 = 0, z_n = G(z_{n-1}) = G_{1,n}(0)$. Then $z_n \uparrow z^* \in [0,1]$, and z^* is the minimal nonnegative solution of equation $z = G(z)$. And it is easy to see that $G(z)$ is an increasing and convex function and $G(1) = 1$. Thus equation $z = G(z)$ has at most two solutions for $z \in (0,1]$. If $\rho = G'(1) > 1$, then $z = G(z)$ has exactly two solutions for $z \in (0,1]$, so that $0 < z^* < 1$. If $\rho = G'(1) \leqslant 1$, then $z = G(z)$ has exactly one solution for $z \in (0,1]$, that is, $z^* = 1$.

But from definition of z_n, we can inductively prove that $z_n = p_{10}^{(n)}$. Indeed,

$$z_0 = 0, \quad z_1 = G(0) = p_{10} = a_0,$$

$$z_n = G(z_{n-1}) = \sum_{k \geqslant 0} a_k \left(p_{10}^{(n-1)} \right)^k = \sum_{k \geqslant 0} p_{1k} p_{k0}^{(n-1)} = p_{10}^{(n)}.$$

Hence $f_{10} = z^*$. $\square$

In the following, we will study a class of extended branching process, that is, when arriving state 0, the process may reach state i in the next step with probability p_i. Assume $\sum_{i=0}^{\infty} p_i = 1$. Namely, the transition probability matrix for the extended branching process is

$$\tilde{p}_{ij} = \begin{cases} p_j, & \text{if } i = 0; \\ p_{ij}, & \text{if } i \neq 0. \end{cases}$$

Here is the main result for the extended branching processes.

Theorem 1.53.
(1) The extended branching process $\tilde{P} = (\tilde{p}_{ij})$ is recurrent iff $\rho \leqslant 1$;
(2) Assume that $p_i = a_i, i \geqslant 0$. When $G'(1) < 1$, the process is positive recurrent;
(3) Assume that $p_i = a_i, i \geqslant 0$. If the extended branching process is ergodic, and its stationary distribution $(\pi_i : i \in E)$ satisfies $m := \sum_{k=0}^{\infty} k\pi_k < \infty$, then $\rho < 1$.

Proof (1) From the proof of Theorem 1.38, we know that the process $\tilde{P}$ is recurrent iff for process P, $f_{i0} = 1, \forall i \geqslant 1$. This is equivalent to $\rho \leqslant 1$.

(2) Note $\tilde{f}_{i0}^{(n)} = f_{i0}^{(n)}$ and $p_{i0}^{(n)} = \sum_{m=1}^{n} f_{i0}^{(m)}$ for $i \neq 0$. Then

$$m_{i0} = \sum_{n=1}^{\infty} \sum_{m \geqslant n} f_{i0}^{(m)} = \sum_{n=1}^{\infty} \left(1 - p_{i0}^{(n-1)} \right).$$

From (1.48), we know that $m_{00} < \infty$ if and only if

$$\sum_{i=1}^{\infty} p_i \sum_{n=1}^{\infty} \left(1 - p_{i0}^{(n)}\right) = \sum_{n=1}^{\infty} \sum_{i=1}^{\infty} a_i \left(1 - p_{i0}^{(n)}\right) = \sum_{n=1}^{\infty} (1 - G(z_{n-1})) < \infty,$$

where $z_n = p_{10}^{(n)} \leqslant 1$. Then by the strict convexity of G on $(0, 1]$, there exists $\delta < 1$ such that $\sup_{z \in [0,1]} G'(z) \leqslant \delta$. Thus

$$1 - G(z) = \int_z^1 G'(z)dz \leqslant \delta(1 - z) \leqslant \delta,$$

$$1 - G(G(z)) = \int_{G(z)}^1 G'(z)dz \leqslant \delta(1 - G(z)) \leqslant \delta^2,$$

$$\cdots\cdots$$

Since $z_0 = 0$ and $z_{n+1} = G(z_n)$, we have

$$\sum_{n=1}^{\infty} (1 - G(z_n)) \leqslant \sum_{n=1}^{\infty} \delta^n < \infty.$$

(3) Set $\Pi(z) = \sum_{k=0}^{\infty} \pi_k z^k$. Then $\Pi'(1) = m$. Since $\pi_j = \sum_i \pi_i \tilde{p}_{ij}$ and $G_{i,1}(z) = G(z)^i$, we have

$$\Pi(z) = \sum_{i,j} \pi_i \tilde{p}_{ij} z^j = \pi_0 \sum_{j=0}^{\infty} a_j z^j + \sum_{i=1}^{\infty} \pi_i \sum_{j=0}^{\infty} p_{ij} s^j$$

$$= \pi_0 G(z) + \Pi(G(z)) - \pi_0.$$

Differentiating in z and letting $z \uparrow 1$, it follows that

$$\rho = G'(1) = \lim_{z \uparrow 1} \frac{\Pi'(z)}{\pi_0 + \Pi'(z)} = \frac{m}{\pi_0 + m} < 1. \quad \square$$

1.6 Supplements and Exercises

(1) Prove that for any measurable sets A, B, C, once $\mathbb{P}(AB) > 0$, we have $\mathbb{P}[BC|A] = \mathbb{P}[B|A]\,\mathbb{P}[C|AB]$.

(2) Definition 1.11 of Markov property has the following equivalent statement: for any $r \geqslant 1$, $\ell_1 < \ell_2 < \cdots < \ell_r < m < n$ and $i_1, \cdots, i_r, i, j, k \in E$, we have

$$\mathbb{P}[X_{\ell_1} = i_1, \cdots, X_{\ell_r} = i_r, X_n = k | X_m = j]$$
$$= \mathbb{P}[X_{\ell_1} = i_1, \cdots, X_{\ell_r} = i_r | X_m = j]\,\mathbb{P}[X_n = k | X_m = j].$$

(3) Prove that the nth power $P^n = \left(p_{ij}^{(n)}\right)$ of Markov P is the n-step transition probability, that is, $p_{ij}^{(n)} = \mathbb{P}[X_n = j | X_0 = i]$.

(4) If $p_{jk}^{(n)} > 0$, then there exists $m \leqslant n$ such that $e_{jk}^{(m)} > 0$.
Hint: Follow the proof of Lemma 1.14. Given n, let

$$\xi_j = \max\{m \leqslant n : X_m = j\}.$$

(5) Prove $\lim\limits_{n \to \infty} p_{ii}^{(n)} = \lim\limits_{s \uparrow 1}(1 - s)P_{ii}(s)$ (apply Abel's Theorem).

(6) Let $(X_n)_{n \geqslant 0}$ be a Markov chain on E. Assume $\phi : E \to \tilde{E}$ is a one-to-one map. Let $\tilde{X}_n = \phi(X_n)$. Then $(\tilde{X}_n)$ is a Markov chain on $\tilde{E}$, and it has the same transition matrix as (X_n). If ϕ is not one-to-one, then what will happen?

(7) Define $\alpha_{ij} = \mathbb{P}_i[N_j = \infty]$, where $N_j = \#\{n \geqslant 1 : X_n = j\}$ is the visit times to state j of the Markov chain.
  (a) Prove $\alpha_{ii} = 1$ iff i is recurrent.
  (b) If j is recurrent, then $\alpha_{ij} = f_{ij}$; if j is transient, then $\alpha_{ij} = 0$.
  Hint: Define $\alpha_{ij}^{(k)} = \mathbb{P}_i[N_j \geqslant k]$ and prove $\alpha_{ij}^{(k)} = f_{ij}\alpha_{jj}^{(k-1)}$.

(8) Assume $(X_n)_{n \geqslant 0}$ and $(Y_n)_{n \geqslant 0}$ are two independent irreducible Markov chains, with periods d_1 and d_2, respectively. Prove that $(Z_n := (X_n, Y_n))_{n \geqslant 0}$ is irreducible if and only if d_1 and d_2 are prime to each other. And in this case, (Z_n) has period $d_1 d_2$.

(9) (Queueing theory) Assume $\{\xi_n : n \geqslant 0\}$ is a sequence of independent and identically distributed random variables taking values on $\mathbb{Z}_+$, with probability distribution $\mathbb{P}[\xi_n = i] = a_i \, (i = 0, 1, 2, \cdots)$. Let $X_{n+1} = (X_n - 1)^+ + \xi_n$. Prove that $(X_n)_{n \geqslant 0}$ is a Markov chain and write out its transition probability matrix.
(This is a discrete-time queueing model. Regard ξ_n as the customers arriving at a service deck, and X_{n+1} as the customers waiting for service at time $n + 1$.)

(10) (Inventory model) Assume $\{\xi_n : n \geqslant 0\}$ is a sequence independent and identically distributed random variables taking values in $\mathbb{Z}_+$, with probability distribution $\mathbb{P}[\xi_n = i] = a_i \, (i = 0, 1, 2, \cdots)$. Fix $0 < s < S < \infty$. Let

$$X_{n+1} = \begin{cases} X_n - \xi_{n+1}, & \text{if } s < X_n \leqslant S; \\ S - \xi_{n+1}, & \text{if } X_n \leqslant s. \end{cases}$$

Prove that $(X_n)_{n \geqslant 0}$ is a Markov chain and write out its transition probability matrix.
(This is a typical inventory model. ξ_n is the quantity for the daily need,

and (s, S) is inventory storage. Namely, when the inventory X_n is not more than s, it should be stocked to the inventory S; when inventory is more than s, it need not be stocked any more.)

(11) (Gene model) Suppose there are two types of genes: one is type A, the other is type B, and the total number of genes is $2N$. The constitution of the next generation gene is determined by the independent binomial trials of $2N$ times. Namely, the parent generation contains i type A gene and $2N - i$ type B, then the result for each trial is type A or type B with probability

$$p_i = \frac{i}{2N}, \qquad q_i = 1 - p_i,$$

respectively. Let X_n be the number of type A gene in nth generation. Prove that $(X_n)_{n \geqslant 0}$ is a Markov chain. Write out its transition probability matrix and prove that

(a) $0, 2N$ are two absorbing states;

(b) $\lim_{n \to \infty} p_{ij}^{(n)} = 0$ for $0 < j < 2N$; while $\lim_{n \to \infty} p_{i0}^{(n)} = 1 - i/2N$, $\lim_{n \to \infty} p_{i,2N}^{(n)} = i/2N$.

(12) (Branching process) Assume that $\{Y_{n,k} : n, k \geqslant 1\}$ is a sequence of independent and identically distributed random variables taking values in $\mathbb{Z}_+$, with distribution $\mathbb{P}[Y_{n,k} = i] = \mu_i, i = 0, 1, 2, \cdots$. Let

$$X_{n+1} = \sum_{k=1}^{X_n} Y_{n+1,k}, \qquad X_0 = 1, \quad n \geqslant 0.$$

(a) Prove that $(X_n)_{n \geqslant 0}$ is a Markov chain, and its transition probability matrix $P = (p_{ij})$ is

$$p_{ij} = \frac{1}{j!} \frac{\partial^j}{\partial x^j} \left(\sum_{k=0}^{\infty} \mu_k x^k \right)^i \Bigg|_{x=0};$$

(b) Let $m = \mathbb{E}X_1, \sigma^2 = \text{Var}(X_1)$ (variance). Then $\mathbb{E}X_n = m^n$,

$$\text{Var}X_n = \begin{cases} \sigma^2 m^{n-1} \frac{m^n - 1}{m - 1}, & \text{if } m \neq 1; \\ n\sigma^2, & \text{if } m = 1. \end{cases}$$

(13) (Markov chains on general state spaces) Let $(E, \mathscr{E})$ be a measurable space, $P(x, A) (A \in \mathscr{E})$ be a transition probability measure. If random process $X = (X_n)_{n \geqslant 0}$ has the following Markov property:

$$\mathbb{P}[X_{n+1} \in A | X_n, X_{n-1}, \cdots, X_0] = P(X_n, A),$$

then X is called a discrete-time Markov process on E. Try to write out the distribution of $(X_0, X_1, \cdots, X_n)$.

(14) Assume $(X_n)_{n\geqslant 0}$ is an irreducible, aperiodic Markov chain on $\mathbb{N}$. Prove the chain is transient if and only if for each initial distribution, it holds $X_n \to \infty$ a.s.; while the chain is null-recurrent if and only if the above convergence is in the sense of probability but not in the sense of a.s..

(15) Prove Corollary 1.21.

(16) (Classical coupling) Let $(X_n)_{n\geqslant 0}$ and $(Y_n)_{n\geqslant 0}$ be two irreducible, aperiodic Markov chains. Define the coupling time

$$\tau(\omega) = \inf\{n \geqslant 0 : X_n(\omega) = Y_n(\omega)\}.$$

Construct the coupling process Z_n as follows:

$$Z_n(\omega) = \begin{cases} (X_n, Y_n)(\omega), & \text{if } n < \tau(\omega); \\ (X_n, X_n)(\omega), & \text{if } n \geqslant \tau(\omega). \end{cases}$$

(a) Prove the coupling process $(Z_n)_{n\geqslant 0}$ is a Markov chain.

(b) Study its irreducibility and periodicity.

(c) If $(X_n)_{n\geqslant 0}$ and $(Y_n)_{n\geqslant 0}$ are (positive) recurrent, then so is $(Z_n)_{n\geqslant 0}$?

(17) Let $(X_n)_{n\geqslant 0}$ be a recurrent Markov chain, and $X_0 = i$. Consider the successive return times to i:

$$\tau_1 = \inf\{n > 0 : X_n = i\}, \quad \tau_{k+1} = \inf\{n > \tau_k : X_n = i\}, k \geqslant 1.$$

(a) Prove $\tau_{k+1} - \tau_k, k \geqslant 1$ are independent and identically distributed.

(b) Let $Y_n = X_{n+\tau_1}$, then $(Y_n)_{n\geqslant 0}$ is a homogeneous Markov chain, and it has the same transition probability matrix as that of $(X_n)_{n\geqslant 0}$.

(18) (Double stochastic matrix) Assume that $P = (p_{ij})$ is a finite irreducible double stochastic matrix, that is, $\sum_{k=1}^{m} p_{kj} = \sum_{k=1}^{m} p_{ik} = 1$ for each $i, j = 1, \cdots, m$. Prove that the stationary distribution for this Markov chain is the uniform distribution on $\{1, 2, \cdots, m\}$.

(19) (Dual chain) Assume that P is a Markov chain with invariant measure (μ_i). Define $\bar{p}_{ij} = \mu_j p_{ji}/\mu_i$. Prove $\bar{P} = (\bar{p}_{ij})$ is a Markov chain. $\bar{P}$ is called the dual to P (with respect to measure μ). And then prove

(a) Markov chain P is irreducible iff so is its dual chain $\bar{P}$;

(b) Markov chain P is recurrent iff so is its dual chain $\bar{P}$;

(c) Markov chain P is positive recurrent iff so is its dual chain $\bar{P}$.

(20) (Reversibility) A Markov chain X_n is called reversible, if for any time $n, m \geqslant 0$ and states $i, j \in E$, we have

$$\mathbb{P}[X_n = i, X_m = j] = \mathbb{P}[X_n = j, X_m = i]. \tag{1.37}$$

Prove the following assertions.

(a) (1.37) is equivalent to for any $n_1 \leqslant n_2 \leqslant \cdots \leqslant n_m$ with $n_2 - n_1 = n_m - n_{m-1}$, $n_3 - n_2 = n_{m-1} - n_{m-2}, \cdots$ and states $i_1, \cdots, i_m \in E$,
$$\mathbb{P}[X_{n_1} = i_1, \cdots, X_{n_m} = i_m] = \mathbb{P}[X_{n_1} = i_m, \cdots, X_{n_m} = i_1].$$

(b) A Markov chain is reversible if and only if there exists stationary distribution (π_i) such that $\pi_i p_{ij} = \pi_j p_{ji}$.

(21) Suppose that the first column of transition probability matrix P is $\{q_0, q_1, \cdots\}$ and $p_{i,i+1} = 1 - q_i$, $i = 0, 1, \cdots$.

(a) When is the chain irreducible?

(b) Prove if the chain is irreducible, then the chain is recurrent if and only if $\sum_i q_i = \infty$.

(c) When is the chain null-recurrent, and positive recurrent?

(22) Assume the first row of transition probability matrix P is $\{q_0, q_1, \cdots\}$, and $p_{i,i-1} = 1$, $i = 1, 2, \cdots$, where $q_i > 0$ and $\sum_i q_i = 1$.

(a) Prove the chain is irreducible and recurrent.

(b) When the chain is null-recurrent or positive recurrent?

(23) Consider a Markov chain on $\mathbb{Z}_+$ with transition probability matrix $p_{0i} = a_i$, $p_{i0} = 1$, where $a_i > 0$ and $\sum_i a_i = 1$. Study its periodicity, recurrence and positive recurrence.

(24) If positive integers $s_1, s_2, \cdots, s_m$ have the greatest common divisor d, then there exists positive integer N, such that when $n > N$, there are nonnegative integers $c_1, c_2, \cdots, c_m$ such that $nd = \sum_{i=1}^{m} c_i s_i$.

Hint: There exist $n_1, \cdots, n_m$ such that $d = n_1 s_1 + \cdots + n_m s_m$. Let $q = \sum_{n_i > 0} n_i s_i, r = -\sum_{n_i < 0} n_i s_i$, then $q = r + d$. If $r \neq 0$, then one can take $N = (r/d)^2$ to make it.

(25) Assume Markov chain P has stationary distribution π. Prove if the initial distribution is π (the distribution of X_0), then for each $n > 0$, the distribution of X_n is π. This is the reason that π is called stationary distribution.

(26) A Markov chain on finite state space has at least one recurrent state, and has not any null-recurrent state. Further, an irreducible Markov chain on finite state space is surely positive recurrent.

(27) Assume that state space C of a Markov chain is irreducible and closed. Then C can be decomposed into d disjoint sets:

$$C = \sum_{m=1}^{d} G_m.$$

Starting from any state in G_m, the next step of the chain surely arrives at some state in $G_{m+1} \pmod d$.

Hint: Firstly, fix a state i, and use Proposition 1.29. Let

$$G_r = \left\{ j \in C : \text{there exists } m \geqslant 0, \text{ such that } p_{ij}^{(md+r)} > 0 \right\}.$$

Then $C = \sum_{m=1}^{d} G_m$ and $G_r \cap G_s = \varnothing \, (r \neq s)$. Secondly, prove that this decomposition is independent of the fixed state i.

(28) Study the classification and the limit distribution $\lim_{n \to \infty} p_{ij}^{(n)}$ of the following Markov chain:

$$P = \begin{pmatrix} \frac{1}{2} & 0 & \frac{1}{2} & 0 \\ 0 & 0 & 0 & 1 \\ \frac{1}{4} & \frac{1}{2} & \frac{1}{4} & 0 \\ 0 & \frac{1}{2} & \frac{1}{2} & 0 \end{pmatrix}.$$

(29) Let

$$P = \begin{pmatrix} 1 - a & a \\ b & 1 - b \end{pmatrix}, \quad 0 < a, b < 1.$$

(a) Prove

$$P^n = \frac{1}{a + b} \begin{pmatrix} b & a \\ b & a \end{pmatrix} + \frac{(1 - a - b)^n}{a + b} \begin{pmatrix} a & -a \\ -b & b \end{pmatrix}.$$

(b) Find $\lim_{n \to \infty} P^n$.

(30) Study the recurrence and positive recurrence of the following Markov chains.

(a) Random walk on the whole line $\mathbb{Z}$: $p_{i,i+1} = p$, $p_{i,i-1} = 1 - p \, (0 < p < 1)$ for each integer i. When $p = 1/2$, it is called a symmetric random walk. Try to determine $p_{00}^{(n)}$.

(b) Random walk on the half line $\mathbb{Z}_+$: $p_{00} = 1 - p$, $p_{i,i+1} = p \, (i \geqslant 0)$, $p_{i,i-1} = 1 - p \, (i \geqslant 1)$.

(c) Markov chain on the whole line $\mathbb{Z}$: $p_{i,i+2} = p, p_{i,i-1} = 1 - p \, (0 < p < 1)$ for each integer i.

(31) (a) Consider the two-dimensional symmetric random walk, that is,

$$p_{(i,j),(i+1,j)} = p_{(i,j),(i-1,j)} = p_{(i,j),(i,j+1)} = p_{(i,j),(i,j-1)} = 1/4$$

for each $(i, j) \in \mathbb{Z}^2$. Then the chain is independent coupling of two one-dimensional symmetric random walks, so that the chain is recurrent.

(b) For the three-dimensional symmetric random walk (that is, from a lattice the chain moves the next step to any of the six nearest neighbors with probability $1/6$), how about the conclusions?

(32) Consider the following Markov chain:

$$P = \begin{pmatrix} p_0 & p_1 & p_2 & \cdots & p_m \\ p_m & p_0 & p_1 & \cdots & p_{m-1} \\ \vdots & \vdots & \vdots & \vdots & \vdots \\ p_1 & p_2 & p_3 & \cdots & p_0 \end{pmatrix},$$

where $0 < p_0 < 1$, $p_0 + p_1 + \cdots + p_m = 1$. Try to determine its stationary distribution $(\pi_i)_{1 \leqslant i \leqslant m}$.

(33) Use the first iteration scheme to prove Propositions 1.47 and 1.48.

(34) Prove the minimal nonnegative solution obtained by the second iteration scheme in Theorem 1.45 is the same as that obtained by the first iteration scheme.

(35) Set $m_{ij} = \mathbb{E}_i \tau_j^+$. For fixed $j \in E$, work out the minimal nonnegative solution of the following equations, and explain their probabilistic meaning.

$$x_i = \sum_{k \neq j} p_{ik} x_k + m_{ij}.$$

(36) Assume that P is irreducible.

(a) Prove that for each fixed $j \in E$, P is recurrent iff the minimal solution of equation

$$x_i = \sum_{k \neq j} p_{ik} x_k + p_{ij}$$

is $x_i^* = 1$ for every $i \in E$.

(b) P is recurrent iff for some (and equivalently, for all) j

$$x_i = \sum_{k \neq j} p_{ik} x_k$$

has only null solution.

(37) Suppose $\tilde{P}$ is defined as in Theorem 1.37. Then

(a) For Markov chains P and $\tilde{P}$, $i \neq 0$, $f_{i0} = \tilde{f}_{i0}$ and $\tilde{p}_{i0}^{(n)} = \sum_{m=1}^{n} f_{i0}^{(m)}$;

(b) Prove

$$\lim_{n \to \infty} \tilde{\eta}_i(n) \leqslant \mathbb{P}_i \left[\limsup_{m \to \infty} \tilde{X}_m = \infty \right] \leqslant 1 - f_{i0}.$$

Thus if P is recurrent, then $\lim_{n \to \infty} \tilde{\eta}_i(n) = 0$.

(38) Suppose P is an ergodic Markov chain on finite state space with stationary distribution $(\pi_i : i \in E)$. Let $D = (d_{ij})$ with $d_{ij} = \sum_{n \geqslant 0} (p_{ij}^{(n)} - \pi_j)$. Then

(a) D is well-defined and finite. Moreover $d_{ii} = \hat{m}_{ii}/m_{ii}^2$ and $d_{ij} = d_{jj} - \pi_j m_{ij}$ for $i \neq j$, where $\hat{m}_{ii} = \frac{1}{2}(\mathbb{E}_i[\tau_i^+]^2 - \mathbb{E}_i\tau_i^+)$;

(b) $\sum_j d_{ij} = 0$ and $\operatorname{tr}(D) = \sum_{j:j\neq i} \pi_j m_{ij} = \sum_{i,j:i\neq j} \pi_i \pi_j m_{ij}$.

(39) Suppose P is an ergodic Markov chain on finite state space, and $\bar{P}$ is its dual chain (see Exercise 19). Then $m_{ii} = \bar{m}_{ii}$ and $\operatorname{tr}(D) = \operatorname{tr}(\bar{D})$. (The term with "$^-$" is corresponding to chain $\bar{P}$.)

(40) According to the argument in the first item in Section 1.5, study the recurrence and ergodicity for random walk on $\mathbb{Z}$. That is, assume $(X_n)_{n\geqslant 0}$ is random walk on a line, with the following transition probability matrix $P = (p_{ij})$. For $i \in \mathbb{Z}$, $p_{i,i+1} = p_i$, $p_{i,i-1} = q_i$, $p_{ii} = r_i$ and $p_i + q_j + r_i = 1$. Assume $p_i > 0, r_i \geqslant 0, q_i > 0$. Obviously, if $r_i = 0$ for every i, the chain has period 2; otherwise, it is aperiodic. Assume $r_0 > 0$, then the chain is aperiodic. Study its recurrence and ergodicity.

(41) In the queueing theory, suppose the customers arriving at the service desk in a unit time interval is distributed as Poisson or geometric distribution, respectively. Study the recurrence and ergodicity of the corresponding chain, and work out the stationary distribution in the ergodic case.

(42) (Dual of queueing theory) Assume $\{\xi_n\}_{n\geqslant 0}$ is a sequence of independent and identically distributed random variables taking values in $\mathbb{Z}_+$ with distribution $\mathbb{P}[\xi_n = i] = \mu_i$ $(i = 0, 1, 2, \cdots)$. Let

$$X_{n+1} - (X_n + 1 - \xi_n)^+, n \geqslant 1.$$

(a) Prove $(X_n)_{n\geqslant 0}$ is a Markov chain, and write out the transition probability matrix;

(b) Denote $\rho = \sum_{k=0}^{\infty} k\mu_k$. Then the chain is recurrent if and only if $\rho \geqslant 1$, while it is ergodic if and only if $\rho > 1$.

(c) When ergodic, how do we get the stationary distribution?

(This is the dual of queuing theory. It means that the arriving customer is 1 in a unit time interval, and the services are stochastic, namely, ξ_n is the number of services in the nth time interval. X_{n+1} is the customers waiting for service in the $(n + 1)$th time interval.)

(43) The transition probability matrix $P = (p_{ij})$ is as follows:

$$p_{ij} = \begin{cases} b_j, & \text{if} \quad i = 0; \\ a_{j-i+1}, & \text{if} \quad j \geqslant i - 1; \\ 0, & \text{else,} \end{cases}$$

where $a_k > 0$, $b_k > 0$, $\sum\limits_{k=0}^{\infty} a_k = 1$, $\sum\limits_{k=0}^{\infty} b_k = 1$. Study the recurrence and ergodicity of P.

(44) (Branching process continued)

(a) Use probabilistic method to give a direct proof of $f_{i0} = [f_{10}]^i$.
    Hint: By independence and definition of f_{i0}, we have $f_{i0} = \lim\limits_{n \to \infty} p_{i0}^{(n)}$.

(b) For branching process, study the behavior of X_n when $n \to \infty$ in the case of $\rho = \sum\limits_{k=1}^{\infty} ka_k > 1$.

(c) For the extended branching process, denote $P(z) = \sum\limits_{k=0}^{\infty} p_k z^k, z \in [0,1]$. In the case of $\rho < 1$, how to obtain its stationary distribution?

(45) Let P be a transition probability matrix. Define

$$V(P) = \frac{1}{2} \sup_{i,j} \sum_k |p_{ik} - p_{jk}|.$$

(a) Prove

$$V(P^{n+m}) \leqslant V(P^n)V(P^m).$$

(b) Use τ defined in Exercise 16 to prove

$$V(P^n) \leqslant \sup_{i,j} \mathbb{P}_{ij}[\tau > n].$$

(c) If

$$\sup_{i,j} \mathbb{E}_{ij}\tau < \infty,$$

then $\lim\limits_{n \to \infty} V(P^n) = 0$.

(46) Assume P is a symmetric Markov chain on finite state space, that is, $p_{ij} = p_{ji}$.

(a) For given $1 \leqslant i \leqslant N$, define

$$E_i(n) = \sum_j p_{ij}^{(n)} \log p_{ij}^{(n)}. \tag{1.38}$$

Prove $E_i(n)$ is a decreasing function in $n \geqslant 0$. Furthermore, there exist $C < \infty$ and $\alpha < 1$ such that

$$\log N + E_i(n) \leqslant C\alpha^n.$$

(b) For function ψ, similar to (1.38), define

$$\Psi_i(n) = \sum_j \psi(p_{ij}^{(n)}).$$

Study for what kind of function ψ, the functional Ψ_i has the monotonicity and convergence as in (a).

(47) For a finite irreducible and aperiodic Markov chain, prove the following assertions:

(a) there exists n such that $p_{ij}^{(n)} > 0$ for every i, j;

(b) there exist $C < \infty$ and $\rho < 1$ such that for every i, j

$$|p_{ij}^{(n)} - \pi_j| \leqslant C\rho^n, \quad n \geqslant 0.$$

Chapter 2

Continuous-Time Markov Chains

2.1 Continuous-Time Markov Chains. Uniqueness

Definition 2.1. Assume E is countable. $(X_t)_{t\geqslant 0}$ is called a **Markov chain**, if for every $n \geqslant 2$, $0 \leqslant t_1 < t_2 < \cdots < t_n$ and any $i_1, i_2, \cdots, i_n \in E$, the following **Markov property** holds:

$$\mathbb{P}\big[X_{t_n} = i_n | X_{t_1} = i_1, \cdots, X_{t_{n-1}} = i_{n-1}\big] = \mathbb{P}\big[X_{t_n} = i_n | X_{t_{n-1}} = i_{n-1}\big].$$

The chain is called (time-)homogeneous, if

$$\mathbb{P}[X_t = j | X_s = i] = \mathbb{P}[X_{t-s} = j | X_0 = i] =: p_{ij}(t - s).$$

Denote by $P(t) = (p_{ij}(t))$, which is called **transition probability matrix**. Clearly, it possesses the following properties:

(1) nonnegativeness $P(t) \geqslant 0$, $t \geqslant 0$;

(2) norm condition $P(t)1 = 1$, i.e. $\sum_j p_{ij}(t) = 1$, $t \geqslant 0$;

(3) Chapman-Kolmogorov equation $P(t + s) = P(t)P(s)$;

(corresponding to Markov property)

additionally, we make the following natural assumption:

(4) continuity condition (jump condition) $\lim_{t \to 0} P(t) = I = (\delta_{ij})$.

As proved in [19; §II.1], the last assumption is more or less equivalent to the Lebesgue measurability of the transition probability $p_{ij}(t)$ on $(0, \infty)$ for each $i, j \in E$. Then $(p_{ij}(t))$ is called a **standard transition matrix** in [19].

In 1936, A. N. Kolmogorov proved that:

$$\lim_{t \downarrow 0} \frac{1 - p_{ii}(t)}{t} =: q_i \in [0, \infty], \quad \lim_{t \downarrow 0} \frac{p_{ij}(t)}{t} =: q_{ij} \in [0, \infty), \quad j \neq i, \quad (2.1)$$

and $\sum_{j \neq i} q_{ij} \leqslant q_i$. The last inequality follows from

$$\sum_{j \neq i} \frac{p_{ij}(t)}{t} = \frac{1 - p_{ii}(t)}{t}.$$

and Fatou's lemma, once the existence of the limits is known. The existence of the limits is far away from trivial. As in (4) above only the continuity is assumed, this is not enough for the differentiability at 0. Indeed, we have used the sub-additive property to prove the existence of the first limit.

Denote by $Q = (q_{ij})$, where $q_{ii} = -q_i$. It is then called a Q-**matrix**. Note that q_i may be ∞, and it may happens that $\sum_{j \neq i} q_{ij} < q_i$. These may cause many difficulties in mathematics, however in practice we seldom meet with so much complicated situations.

Definition 2.2. A Q-matrix is called **totally stable** if for all i, $q_i < \infty$; it is called **conservative** if for all i, $\sum_{j \neq i} q_{ij} = q_i$, or $\sum_j q_{ij} = 0$.

In the totally stable case, Condition (4) implies that the trajectories of the process are almost surely (right continuous) step functions. For this reason, we call it **jump condition**.

Theorem 2.3 (Probabilistic meaning of Q-matrix).

(1) $\mathbb{P}_i[X_s = i, 0 \leqslant s \leqslant t] = e^{-q_i t}$. Therefore, if $q_i = 0$, then $\mathbb{P}_i[X_s = i, \forall s \geqslant 0] = 1$, i.e. i is an absorbing state.

(2) Assume that $q_i \in (0, \infty)$. Let $\tau_1 = \inf\{t \geqslant 0 : X_t \neq X_0\}$ be the first jump of (X_t). Then $\mathbb{P}_i[X_{\tau_1} = j] = q_{ij}/q_i$ for $j \neq i$.

Proof (1) Since the trajectories are right-continuous, by Markov property, we have

$$\mathbb{P}_i[X_s = i, 0 \leqslant s \leqslant t] = \lim_{n \to \infty} \mathbb{P}_i[X_{kt/2^n} = i, k = 0, 1, \cdots, 2^n]$$

$$= \lim_{n \to \infty} p_{ii}(t/2^n)^{2^n}$$

$$= \lim_{n \to \infty} \left(1 - \frac{q_i t}{2^n} + o(2^{-n})\right)^{2^n} \qquad \text{(by (2.1))}$$

$$= e^{-q_i t}.$$

(2) Let $i \neq j$. Then

$$\mathbb{P}_i[X_{\tau_1} = j] = \lim_{h \to 0} \mathbb{P}[X_{t+h} = j | X_t = i, X_{t+h} \neq i]$$

$$= \lim_{h \to 0} \frac{\mathbb{P}[X_{t+h} = j | X_t = i]}{\mathbb{P}[X_{t+h} \neq i | X_t = i]}$$

$$= \lim_{h \to 0} \frac{p_{ij}(h)/h}{(1 - p_{ii}(h))/h}$$

$$= \frac{q_{ij}}{q_i}. \qquad \square$$

The following picture illustrates the sample path for the continuous-time Markov chain.

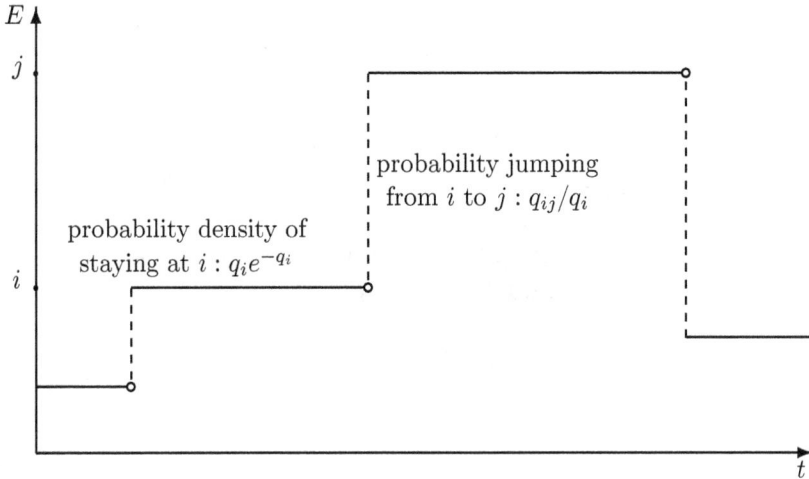

From now on, we always assume that Q-matrix is totally stable and conservative.

Starting from Chapman-Kolmogorov equation $P(t + s) = P(t)P(s)$, making the differentiation at 0 on t and s, respectively, we deduce formally two differential equation (Kolmogorov (1931)):

$$\text{Kolmogorov backward equation} \quad P'(t) = QP(t), \qquad (2.2)$$

$$\text{Kolmogorov forward equation} \quad P'(t) = P(t)Q. \qquad (2.3)$$

Two equations are infinite-dimensional differential equations. Alike in the theory of ordinary differential equations, in practice we know Q rather than $P(t)$.

Definition 2.4. Let $P(t)$ satisfy Conditions (1), (3), (4) in Definition 2.1, and $P(t)1 \leqslant 1$. Then $P(t)$ is called a Q-**process**, if $P'(t)|_{t=0} = Q$.

Now, a natural question is that: for a given Q-matrix, does Q-process exist? Is it unique? First of all, we have

Theorem 2.5. Each Q-process satisfies the Kolmogorov backward equation.

The proof for this theorem is not so difficult. Starting from this, the existence and uniqueness for Q-process are transferred to the existence and uniqueness of the solution to the Kolmogorov backward equation. However, Q-process does not always satisfy the Kolmogorov forward equation. In other words, these two equations are not in the same situation. Before moving forward, we should explain the equivalent integral equations:

backward equation

$$p_{ij}(t) = \sum_{k \neq i} \int_0^t q_{ik} e^{-q_i(t-s)} p_{kj}(s) \mathrm{d}s + \delta_{ij} e^{-q_i t} \qquad (2.4)$$

and forward equation

$$p_{ij}(t) = \sum_{k \neq j} \int_0^t p_{ik}(s) q_{kj} e^{-q_j(t-s)} \mathrm{d}s + \delta_{ij}^{-q_j t}. \qquad (2.5)$$

These two integral equations have good probabilistic interpretation. The former uses the decomposition by the first jump, while the latter uses the decomposition by the last jump. Let us give the details.

Assume $i \neq j$. Note that $e^{-q_i t} = \mathbb{P}_i[X_s$ does not jump during $[0, t]]$. If $q_i \neq 0$, then the process must jump with positive probability during $[0, t]$, so that

$$p_{ij}(t) = \int_0^t q_i e^{-q_i s} \sum_{k \neq i} \frac{q_{ik}}{q_i} p_{kj}(t - s) \mathrm{d}s,$$

where $q_i e^{-q_i s}$ is the probability density for the process jumping at time s (the probability that the process jumps in interval $(s, s + \Delta s)$ is $e^{-q_i s}(1 - e^{-q_i \Delta s})$), q_{ik}/q_i is the probability that it jumps from i to k; $p_{kj}(t - s)$ denotes the probability that starting from k, the process arrives at j after $t - s$. Replacing $t - s$ by s, we get equation (2.4).

Secondly, assume again that $i \neq j$. Then

$$p_{ij}(t) = \sum_{k \neq j} \int_0^t \mathbb{P}_i[X_s = k, X_{s+\mathrm{d}s} = j, \text{ no jump in } [s + \mathrm{d}s, t]]$$

$$= \sum_{k \neq j} \int_0^t \mathbb{P}_i[\text{no jump in } [s + \mathrm{d}s, t] | X_s = k, X_{s+\mathrm{d}s} = j]$$

$$\cdot \mathbb{P}_i[X_{s+\mathrm{d}s} = j | X_s = k] \cdot \mathbb{P}_i[X_s = k]$$

$$= \sum_{k \neq j} \int_0^t e^{-q_j(t-s)} p_{kj}(\mathrm{d}s) p_{ik}(s)$$

$$= \sum_{k \neq j} \int_0^t e^{-q_j(t-s)} q_{kj} p_{ik}(s) \mathrm{d}s. \quad (\text{since } p_{kj}(\mathrm{d}s) = q_{kj}\mathrm{d}s)$$

From this, we deduce the integral form of the forward equation (2.5).

Now, we have the first main result.

Theorem 2.6 ([23]). *Q*-process always exists. Namely, the backward equation and the forward equation admit the same minimal solution $P^{\min}(t)$. Therefore for any *Q*-process, $p_{ij}(t) \geqslant p_{ij}^{\min}(t)$.

Indeed, using the first iteration scheme in Theorem 1.44, we have

$$p_{ij}^{(0)}(t) = 0,$$

$$p_{ij}^{(n+1)}(t) = \sum_{k \neq i} \int_0^t e^{-q_i(t-s)} q_{ik} p_{kj}^{(n)}(s) \mathrm{d}s + \delta_{ij} e^{-q_i t}, \quad n \geqslant 0.$$

Then $p_{ij}^{(n)}(t) \uparrow p_{ij}^{\min}(t)$.

In the study of Q-process, we often use a simpler method, the Laplace transform as in the study of ordinary differential equations. Set

$$p_{ij}(\lambda) = \int_0^\infty e^{-\lambda t} p_{ij}(t) \mathrm{d}t, \quad \lambda > 0,$$

which is called **resolvent**. Now, four conditions for $P(t)$ plus the condition $P'(t)\big|_{t=0} = Q$ are transferred to the following five conditions:
  (1) nonnegativeness $P(\lambda) \geqslant 0$;
  (2) norm condition $\lambda P(\lambda)1 \leqslant 1$;
  (3) resolvent condition $P(\lambda) - P(\mu) + (\lambda - \mu)P(\lambda)P(\mu) = 0$;
  (4) jump condition $\lim\limits_{\lambda \to \infty} (\lambda P(\lambda) - I) = 0$;
  (5) Q-condition $\lim\limits_{\lambda \to \infty} \lambda(\lambda P(\lambda) - I) = Q$.
Furthermore, two integral equations are transferred to the following algebraic equations:

$$\text{Backward equation} \quad p_{ij}(\lambda) = \sum_{k \neq i} \frac{q_{ik}}{\lambda + q_i} p_{kj}(\lambda) + \frac{\delta_{ij}}{\lambda + q_i}, \qquad (2.6)$$

$$\text{Forward equation} \quad p_{ij}(\lambda) = \sum_{k \neq j} p_{ik}(\lambda) \frac{q_{kj}}{\lambda + q_j} + \frac{\delta_{ij}}{\lambda + q_j}. \qquad (2.7)$$

For these two equations, we can obtain the same minimal solution $P^{\min}(\lambda)$, and each Q-process satisfies $P(\lambda) \geqslant P^{\min}(\lambda)$.

Now, we have

$$0 \leqslant p_{ij}(\lambda) - p_{ij}^{\min}(\lambda) = \sum_{k \neq i} \frac{q_{ik}}{\lambda + q_i} \left[p_{kj}(\lambda) - p_{kj}^{\min}(\lambda) \right].$$

If Q-process is not unique, then equation

$$u_i = \sum_{k \neq i} \frac{q_{ik}}{\lambda + q_i} u_k$$

has non-null and nonnegative bounded solution. That is, equation

$$\begin{cases} (\lambda + q_i)u_i = \sum_{k \neq i} q_{ik} u_k, \\ 0 \leqslant u_i \leqslant 1 \end{cases} \qquad (2.8)$$

has non-trivial solution. From this, we can deduce easily the following criterion for uniqueness:

Theorem 2.7 ([24; 55]). Q-process is unique iff equation (2.8) has only the null solution for some (equivalently, all) $\lambda > 0$.

The difficulty to prove of this theorem is of course the necessity. That is, when (2.8) has non-null solutions, how do we construct all Q-processes other than $P^{\min}(\lambda)$:

$$P(\lambda) = P^{\min}(\lambda) + ?$$

Actually, we can construct infinitely many processes. Therefore, when non-unique, there exist infinitely many Q-processes, this, in turn, reflects the importance of uniqueness.

The above criterion has many applications, of which is the following simplest one.

Theorem 2.8. If $M := \sup_i q_i < \infty$, then Q-process is unique.

Proof Suppose that u_i is a solution for equation (2.8). Let $\bar{u} = \sup_i u_i \leqslant 1$. If $\bar{u} > 0$, then for any $\lambda > 0$, we have

$$\bar{u} = \sup_i u_i = \sup_i \sum_{k \neq i} \frac{q_{ik}}{\lambda + q_i} u_k \leqslant \left(\sup_i \sum_{k \neq i} \frac{q_{ik}}{\lambda + q_i} \right) \bar{u}$$

$$= \bar{u} \sup_i \frac{q_i}{\lambda + q_i} \leqslant \bar{u} \frac{M}{\lambda + M} < \bar{u},$$

which is a contradiction. □

Theorem 2.7 has a nice application to the so-called **single birth process**, which will be studied later. The beauty of this theorem is that it relies only on Q-matrix, and we have iteration algorithm of the maximum solution to (2.8). However, it does not sound so good as imagination in practice. Actually, for the multi-dimensional models, equation (2.8) is nearly unsolvable. For some statistical physics models, we struggled for many years to find the answers.

The jump process $P(t) = (p_{ij}(t))$ is called **honest**, if for any $i \in E$, $\sum_{j \in E} p_{ij}(t) = 1$ for every $t \geqslant 0$, or equivalently, for each $i \in E$, $\lambda \sum_{j \in E} p_{ij}(\lambda) = 1$ for some $\lambda > 0$. It is easy to see that the Q-process is unique iff the minimal process $P^{\min}(t)$ (or $P^{\min}(\lambda)$) is honest. The following is a simple but rather practical criterion for uniqueness.

Theorem 2.9 ([4; 61]). Let Q be totally stable and conservative. Then the Q-process is unique iff there exists a sequence of subsets $\{E_n\}$, $E_n \subset E$, and a function $\phi \geqslant 0$ satisfying

(1) $E_n \uparrow E$, $\sup_{i \in E_n} q_i < \infty$, $\lim_{n \to \infty} \inf_{i \notin E_n} \phi_i = \infty$;

(2) there exists $c \in \mathbb{R}$ such that $\sum_{j \in E} q_{ij}\phi_j \leqslant c\phi_i$ for each $i \in E$.

The sequence $\{E_n\}$ is often easier to select and Condition (2) is essential.

Proof We prove the sufficiency first.

a) Take

$$q_{ij}^{(n)} = q_{ij} I_{E_n}(i).$$

From Condition (1), we have $\sup_i q_i^{(n)} < \infty$. From Theorem 2.8 it follows

that $Q^{(n)} = (q_{ij}^{(n)})$ determines uniquely the Q-process $P_n(\lambda) = \left(p_{ij}^{(n)}(\lambda)\right)$.

Next, replacing c by $c^+ = c \vee 0$ if necessary, we see that $Q^{(n)}$ satisfies

Condition (2) with the same function ϕ and c^+. Since $(p_{ij}^{(n)}(\lambda) : i \in E)$ is

the minimal solution to the backward equation

$$x_i = \sum_{k \neq i} \frac{q_{ik}^{(n)}}{\lambda + q_i^{(n)}} x_k + \frac{\delta_{ij}}{\lambda + q_i^{(n)}}, \quad i \in E.$$

From the linear combination theorem (Theorem 1.43), we know that

$$\left(\sum_j p_{ij}^{(n)}(\lambda)\phi_j : i \in E\right)$$

is the minimal solution to

$$x_i = \sum_{k \neq i} \frac{q_{ik}^{(n)}}{\lambda + q_i^{(n)}} x_k + \frac{\phi_i}{\lambda + q_i^{(n)}}.$$

It follows from Condition (2) that

$$\frac{\phi_i}{\lambda - c^+} \geqslant \sum_{k \neq i} \frac{q_{ik}^{(n)}}{\lambda + q_i^{(n)}} \cdot \frac{\phi_k}{\lambda - c^+} + \frac{\phi_i}{\lambda + q_i^{(n)}}, \quad \lambda > c^+$$

$$\Longleftrightarrow \left(\lambda + q_i^{(n)}\right)\phi_i \geqslant \sum_{k \neq i} q_{ik}^{(n)}\phi_k + \phi_i(\lambda - c^+).$$

Then the comparison theorem (Theorem 1.42) implies that

$$\sum_j p_{ij}^{(n)}(\lambda)\phi_j \leqslant \frac{\phi_i}{\lambda - c^+} < \infty, \quad \lambda > c^+.$$

b) When $i \in E_n$, we have

$$p_{iA}^{\min}(\lambda) := \sum_{j \in A} p_{ij}^{\min}(\lambda) = \sum_{k \neq i} \frac{q_{ik}}{\lambda + q_i} p_{kA}^{\min}(\lambda) + \frac{\delta_{iA}}{\lambda + q_i}$$

$$= \sum_{k \neq i} \frac{q_{ik}^{(n)}}{\lambda + q_i^{(n)}} p_{kA}^{\min}(\lambda) + \frac{\delta_{iA}}{\lambda + q_i^{(n)}}$$

for every $A \subset E_n$. And when $i \notin E_n$, we have

$$p_{iA}^{\min}(\lambda) \geqslant 0 = \sum_{k \neq i} \frac{q_{ik}^{(n)}}{\lambda + q_i^{(n)}} p_{kA}^{\min}(\lambda) + \frac{\delta_{iA}}{\lambda + q_i^{(n)}}, \qquad A \subset E_n.$$

Thus the comparison theorem gives

$$p_{iA}^{\min}(\lambda) \geqslant p_{iA}^{(n)}(\lambda), \qquad i \in E, \ A \subset E_n.$$

c) Finally, since $\left(p_{ij}^{(n)}(\lambda)\right)$ is honest, we have

$$\lambda p_{iE_n}^{\min}(\lambda) \geqslant \lambda p_{iE_n}^{(n)}(\lambda) = 1 - \lambda p_{iE_n^c}^{(n)}(\lambda)$$

$$\geqslant 1 - \lambda \sum_{j \notin E_n} p_{ij}^{(n)}(\lambda)\phi_j \Big/ \inf_{i \notin E_n} \phi_i$$

$$\geqslant 1 - \frac{\lambda \phi_i}{\inf_{i \notin E_n} \phi_i} \cdot \frac{1}{\lambda - c_+}, \qquad \lambda > c_+.$$

By letting $n \to \infty$, we obtain that $\lambda p_{iE}^{\min}(\lambda) \geqslant 1$. By the definition of $p_{ij}(\lambda)$, we know that the minimal process is honest, hence the Q-process is unique.

Next, we go to prove the necessity. For this, on an extended space $E_\Delta = E \bigcup \{\Delta\}$ with Δ an extra state outside E, define a probability transition matrix P^Δ by

$$p_{ij}^\Delta = \begin{cases} q_{ij}(1 - \delta_{ij})/(\lambda + q_i), & \text{if} \quad i, j \in E; \\ \lambda/(\lambda + q_i), & \text{if} \quad i \in E, j = \Delta; \\ p_j, & \text{if } i = \Delta, j \in E, \end{cases} \qquad (2.9)$$

where $\lambda > 0$ and $p_i > 0$, $\sum_{i \in E} p_i = 1$.

To prove the necessity, we need the following lemma.

Lemma 2.10. The Q-process is unique iff the Markov chain P^Δ is recurrent.

Now suppose the Q-process is unique. From Lemma 2.10, we see that P^Δ is recurrent. Relabel E as $\mathbb{Z}_+$ for a moment. By Theorem 1.37, there exist $0 \leqslant y_i < \infty (i \in E_\Delta)$ such that $\lim_{i \to \infty} y_i = \infty$ and

$$\sum_{j \in E_\Delta} p_{ij}^\Delta y_j \leqslant y_i, \quad i \in E.$$

Hence, for $i \in E$,

$$(\lambda + q_i)y_i \geqslant \sum_{j \neq i} q_{ij}y_j + \lambda y_\Delta \geqslant \sum_{j \neq i} q_{ij}y_j.$$

By setting $E_n = \{i \in E : y_i \leqslant n\}$, $\phi_i = y_i$, and $c = \lambda$, we have $\sum_{j \in E} q_{ij}\phi_j \leqslant c\phi_i$. Since $0 \leqslant y_i < \infty$ and $\lim_{i \to \infty} y_i = \infty$, it is easy to see that the sequence

$\{E_n, n \geqslant 1\}$ satisfies Condition (1) of our Theorem. Returning to the original E Condition (2) still holds for the same $\{E_n\}, \phi$ and c, except $\lim\limits_{i\to\infty} \phi_i$ is replaced by $\lim\limits_{n\to\infty} \inf\limits_{i\notin E_n} \phi_i$. $\square$

Proof of Lemma 2.10 First note that P^Δ is irreducible. The recurrence of P^Δ is equivalent to that $\hat{f}_{i\Delta} = \sum\limits_{n=1}^{\infty} \hat{f}_{i\Delta}^{(n)} = 1$ for any $i \in E$, where $\hat{f}_{i\Delta}^{(n)}$ is the probability of the first return to Δ at the nth step for the Markov chain P^Δ.

Let $A = (a_{ij})_{ij\in E}$ be the submatrix of P^Δ restricted on $E \times E$, that is, $a_{ij} = q_{ij}(1 - \delta_{ij})/(\lambda + q_i), i, j \in E$. Let $P^{\min}(\lambda) = (p_{ij}^{\min}(\lambda))_{ij\in E}$ be the minimal nonnegative solution of Backward Equation:

$$P(\lambda) = AP(\lambda) + \operatorname{diag}(1/(\lambda + q)),$$

where $\operatorname{diag}(1/(\lambda+q)) = (\delta_{ij}/(\lambda+q_i))_{i,j\in E}$. Use the second iteration scheme for the nonnegative minimal solution (Theorem 1.45):

$$P^n(\lambda) = A^{n-1}\operatorname{diag}(1/(\lambda + q)), \quad n \geqslant 1$$

to derive that

$$P^{\min}(\lambda) = \sum_{n=0}^{\infty} A^n \operatorname{diag}(1/(\lambda + q)).$$

Hence

$$\lambda \sum_{j\in E} p_{ij}^{\min}(\lambda) = \sum_{n\geqslant 0} \sum_{j\in E} a_{ij}^{(n)} \frac{\lambda}{\lambda + q_j}.$$

Since $p_{j\Delta}^\Delta = \lambda/(\lambda + q_j)$ for $j \in E$, it is easy to check that for $n \geqslant 0$

$$\sum_{j\in E} a_{ij}^{(n)} \frac{\lambda}{\lambda + q_j} = \hat{f}_{i\Delta}^{(n+1)}, \quad \forall i \in E.$$

Therefore, for any $i \in E$,

$$\lambda \sum_{j\in E} p_{ij}^{\min}(\lambda) = \sum_{n\geqslant 0} \hat{f}_{i\Delta}^{(n+1)} = \hat{f}_{i\Delta}. \quad \square$$

2.2 Recurrence and Ergodicity

After solving the problem of uniqueness for Q-process, our next task is to study the recurrence and ergodicity for Q-process.

First of all, we consider the classification of the state space. For continuous-time Markov chains, we can prove that

$$p_{ii}(t) > 0, \quad t \geqslant 0; \quad \text{(Easy!)}$$

$p_{ij}(t)$ is either positive or null for all t on $(0, \infty)$. (Difficult!)

Therefore there does not exist the periodicity problem, and we have the following result on classification of the state space. For the definition of an irreducible matrix, see Section 1.1.

Proposition 2.11. $(p_{ij}(t))$ is irreducible iff $Q = (q_{ij})$ is irreducible.

From now on, we restrict ourselves to the irreducible case, and always assume that Q-matrix is **regular**, i.e. it is totally stable, conservative and that Q-process is unique.

Definition 2.12.
(1) $P(t)$ is called **recurrent**, if for each $h > 0$, the discrete-time chain $P(h)$ is recurrent $\Longleftrightarrow \int_0^\infty p_{ii}(t)\mathrm{d}t = \infty$.
(2) $P(t)$ is called **ergodic**, if for each $h > 0$, the discrete-time chain with transition probability matrix $P(h)$ is ergodic $\Longleftrightarrow \lim_{t\to\infty} p_{jj}(t) = \pi_j > 0$.

For the continuous-time Markov chain, the recurrence is easy to deal with. Indeed, the recurrence of a continuous-time Markov chain is equivalent to that of an embedded discrete-time Markov chain. For this, we assume $q_i > 0$ for every i and let

$$\bar{p}_{ij} = q_{ij}/q_i, \quad j \neq i; \qquad \bar{p}_{ii} = 0, \quad i, j \in E.$$

Then $\bar{P} = (\bar{p}_{ij})$ is transition probability matrix. The discrete-time Markov chain with transition probability matrix $\bar{P}$ is called **embedded chain** or **jump chain**. The reason for jump chain is as follows. Let $\tau_1, \tau_2, \cdots$ be the successive jump epochs, that is, let $\tau_1 = \inf\{t \geqslant 0 : X_t \neq X_0\}$ be the first jump epoch and $\tau_n = \inf\{t > \tau_{n-1} : X_t \neq X_{t-}\}$ be the nth jump epoch. Define $Y_n = X_{\tau_n}$. Clearly (Y_n) is a discrete-time Markov chain with transition probability matrix $\bar{P} = (\bar{p}_{ij})$.

Theorem 2.13 ([5]).

$$\int_0^\infty p_{ij}^{\min}(t)\mathrm{d}t = \frac{1}{q_j} \sum_{n=0}^\infty \bar{p}_{ij}^{(n)}. \tag{2.10}$$

Thus

$$\int_0^\infty p_{ii}(t)\mathrm{d}t = \frac{1}{q_i} \sum_{n=0}^\infty \bar{p}_{ii}^{(n)}, \tag{2.11}$$

once the process is unique.

Proof Note for fixed j, $(p_{ij}^{\min}(\lambda) : i \in E)$ is the minimal solution of equation

$$x_i = \sum_{k \neq i} q_{ik} x_k/(\lambda + q_i) + \delta_{ij}/(\lambda + q_i), \qquad i \in E.$$

Using the second iteration scheme for the minimal solutions (Theorem 1.45), we have

$$p_{ij}^{\min}(\lambda) = \sum_{n=0}^{\infty} \bar{Q}(\lambda)^n (i,j)/(\lambda + q_j),$$

where $\bar{Q}(\lambda)^n$ is the nth power of the following matrix:

$$\bar{q}_{ij}(\lambda) = q_{ij}/(\lambda + q_i), \quad i \neq j; \quad \bar{q}_{ii}(\lambda) = 0, \quad i, j \in E.$$

By letting $\lambda \downarrow 0$, the monotone convergence theorem gives that

$$\int_0^\infty p_{ij}^{\min}(t)dt = \lim_{\lambda \to 0} p_{ij}^{\min}(\lambda) = \frac{1}{q_j}\delta_{ij} + \frac{1}{q_j}\sum_{n=1}^{\infty} \bar{p}_{ij}^{(n)}.$$

This proves the assertion.   □

Unfortunately, this equivalence does not remain true for ergodicity between the continuous-time Markov chain and its jump chain. The reader is urged to think about the reason behind this! By using the discrete-time Markov chain $P(h)$, we can easily deduce the following theorem.

Theorem 2.14.
(1) The limits $\lim_{t \to \infty} p_{ij}(t) = \pi_j$ exist, independent of i.
(2) The process is positive recurrent iff the following equation has the unique solution:

$$\pi_i > 0, \quad \sum_i \pi_i = 1, \quad \sum_i \pi_i q_{ij} = 0, \quad i, j \in E.$$

However, we still have the following criterion, similar to the discrete-time case.

Theorem 2.15 ([56; 33; 64]). Suppose Q is regular and irreducible. Let H be a non-empty and finite subset of E. Then the chain is ergodic iff equation

$$\begin{cases} \sum_j q_{ij}y_j \leqslant -1, & i \notin H, \\ \sum_{i \in H}\sum_{j \neq i} q_{ij}y_j < \infty \end{cases} \tag{2.12}$$

has a finite nonnegative solution.

The chain is called **exponentially ergodic**, if there exist constant $\alpha > 0$ and $c_{ij} < \infty$ such that $|p_{ij}(t) - \pi_j| \leqslant c_{ij}e^{-\alpha t}$ holds for every $i, j \in E$ and $t \geqslant 0$. The optimal (greatest) α is called the **exponential ergodicity rate**. We present here the criterion for the exponential ergodicity without proof.

Theorem 2.16 ([64]). Let Q and H be as above. Then the chain is exponentially ergodic iff there exists $0 < \lambda < q_i$ for each i such that equation

$$\begin{cases} \sum_j q_{ij} y_j \leqslant -\lambda y_i - 1, & i \notin H, \\ \sum_{i \in H} \sum_{j \neq i} q_{ij} y_j < \infty \end{cases} \tag{2.13}$$

has finite nonnegative solution.

To prove Theorem 2.15, the key is to establish the recursive relations. Similar to f_{ij} and m_{ij} in the discrete-time case, we define

$$\sigma_H = \inf\{t \geqslant \tau_1 : X_t \in H\}, \quad f_{iH}^{(n)}(t) = \mathbb{P}_i[\sigma_H = \tau_n > t],$$

$$f_{iH}(t) = \sum_{n=1}^{\infty} f_{iH}^{(n)}(t), \quad f_{iH}^{(n)} = f_{iH}^{(n)}(0), \quad f_{iH} = f_{iH}(0). \tag{2.14}$$

Lemma 2.17. Let $\bar{p}_{iH} = \sum_{j \in H} \bar{p}_{ij}$. Then $f_{iH}^{(1)} = \bar{p}_{iH}$, $f_{iH}^{(n+1)} = \sum_{k \notin H} \bar{p}_{ik} f_{kH}^{(n)}$. Furthermore, $(f_{iH} : i \in E)$ is the minimal solution of equation

$$x_i = \sum_{k \notin H} \bar{p}_{ik} x_k + \bar{p}_{iH}. \tag{2.15}$$

If H is a non-empty and finite set, then the chain is recurrent iff $f_{iH} = 1$ for all i.

Proof Note

$$f_{iH}^{(1)} = \mathbb{P}_i[\sigma_H = \tau_1] = \bar{p}_{iH},$$

and

$$f_{iH}^{(n+1)} = \mathbb{P}_i[\sigma_H = \tau_{n+1}] = \sum_{k \notin H} \bar{p}_{ik} \mathbb{P}_k[\sigma_H = \tau_n] = \sum_{k \notin H} \bar{p}_{ik} f_{kH}^{(n)}, \quad n \geqslant 1.$$

Then the first assertion comes from the second iteration scheme, and the second assertion follows from Theorem 2.13, Proposition 1.47 and Lemma 1.46. $\quad\square$

Lemma 2.18. We have

$$f_{iH}^{(1)}(t) = \bar{p}_{iH} e^{-q_i t},$$

$$f_{iH}^{(n+1)}(t) = \int_0^t q_i e^{-q_i s} \sum_{k \notin H} \bar{p}_{ik} f_{kH}^{(n)}(t - s) \mathrm{d}s + \sum_{k \notin H} \bar{p}_{ik} f_{kH}^{(n)} e^{-q_i t}, \quad n \geqslant 1.$$

$$\tag{2.16}$$

Proof Clearly, by Theorem 2.3,

$$f_{iH}^{(1)}(t) = \mathbb{P}_i\,[\,\sigma_H = \tau_1 > t\,] = \bar{p}_{iH}e^{-q_i t}.$$

Next,

$$
\begin{aligned}
f_{iH}^{(n+1)}(t) &= \mathbb{P}_i\,[\,\sigma_H = \tau_{n+1} > t\,]\\
&= \mathbb{P}_i\,[\,\sigma_H = \tau_{n+1} > t \geqslant \tau_1\,] + \mathbb{P}_i\,[\,\sigma_H = \tau_{n+1}, \tau_1 > t\,]\\
&=: \mathrm{I} + \mathrm{II}, \qquad n \geqslant 1.
\end{aligned}
$$

By the strong Markov property (cf. Theorem 4.16 in Chapter 4), we have

$$
\begin{aligned}
\mathrm{I} &= \int_0^t q_i e^{-q_i s} \sum_{k \notin H} \bar{p}_{ik}\mathbb{P}_k\,[\,\sigma_H = \tau_n > t - s\,]\mathrm{d}s\\
&= \int_0^t q_i e^{-q_i s} \sum_{k \notin H} \bar{p}_{ik} f_{kH}^{(n)}(t - s)\mathrm{d}s
\end{aligned}
$$

and

$$
\mathrm{II} = \int_t^\infty q_i e^{-q_i s} \sum_{k \notin H} \bar{p}_{ik}\mathbb{P}_k\,[\,\sigma_H = \tau_n\,]\mathrm{d}s = \sum_{k \notin H} \bar{p}_{ik} f_{kH}^{(n)} e^{-q_i t}.
$$

Finally we obtain the desired result by induction.   □

Remark 2.19. Here we use the strong Markov property for Q-processes, which will be studied in Chapter 4. By Theorem 4.16 there, we know that Markov chains on discrete state spaces always have the strong Markov property. In the following proof of (2.19), we also use the strong Markov property. Roughly speaking, when starting from a random time, the process behaves afterwards like that starting from a deterministic time.

Set

$$
\begin{aligned}
e_{iH}^{(n)}(\lambda) &= \int_0^\infty e^{\lambda t} f_{iH}^{(n)}(t)\mathrm{d}t,\\
e_{iH}(\lambda) &= \sum_{n=1}^\infty e_{iH}^{(n)}(\lambda), \quad 0 \leqslant \lambda < q_i, \quad \forall i \in E.
\end{aligned}
\tag{2.17}
$$

Lemma 2.20. We have

$$
e_{iH}^{(1)}(\lambda) = \frac{1}{q_i - \lambda}\bar{p}_{iH},
$$

$$
e_{iH}^{(n+1)}(\lambda) = \frac{q_i}{q_i - \lambda} \sum_{k \notin H} \bar{p}_{ik} e_{kH}^{(n)}(\lambda) + \frac{1}{q_i - \lambda} \sum_{k \notin H} \bar{p}_{ik} f_{kH}^{(n)}.
$$

In particular, $(e_{iH}(\lambda) : i \in E)$ is the minimal solution to

$$
x_i = \frac{q_i}{q_i - \lambda} \sum_{k \notin H} \bar{p}_{ik}x_k + \frac{1}{q_i - \lambda}f_{iH}, \quad i \in E.
\tag{2.18}
$$

Proof We prove only the latter assertion. For this, by applying Theorem 1.45 to

$$g_i^{(1)} = \bar{p}_{iH}, \qquad g_i^{(n+1)} = \sum_{k \notin H} \bar{p}_{ik} f_{kH}^{(n)}, \qquad g_i = \sum_{n=1}^{\infty} g_i^{(n)},$$

and by Lemma 2.17, we have $g_i = f_{iH}$ $(i \in E)$.   □

Lemma 2.21. If the chain is ergodic, then $e_{iH}(0) = \mathbb{E}_i \sigma_H < \infty$ for each $i \in H$. Furthermore $e_{iH}(0) < \infty$ for each $i \in E$.

Proof a) Set $\sigma_i := \sigma_{\{i\}}$. Then $\sigma_H \leqslant \sigma_i$ for $i \in H$. Let $F_{ij}(t) = \mathbb{P}_i[\sigma_j \leqslant t]$. Similar to the proof of (2.4), we can prove that

$$p_{ij}(t) = \delta_{ij} e^{-q_i t} + \int_0^t p_{jj}(t - s) \mathrm{d}F_{ij}(s). \tag{2.19}$$

By making Laplace transform on both sides of (2.19), we have (for $j = i$)

$$p_{ii}(\lambda) = \frac{1}{\lambda + q_i} + p_{ii}(\lambda) f_{ii}(\lambda),$$

or

$$\lambda p_{ii}(\lambda) = \frac{1}{\lambda + q_i} \left(\frac{1 - f_{ii}(\lambda)}{\lambda} \right)^{-1},$$

where $f_{ii}(\lambda) = \int_0^\infty e^{-\lambda s} \mathrm{d}F_{ii}(s)$. It follows that

$$\lim_{\lambda \to 0} \frac{1 - f_{ii}(\lambda)}{\lambda} = \int_0^\infty s \mathrm{d}F_{ii}(s) = \mathbb{E}_i \sigma_i.$$

Thus

$$0 < \pi_i = \lim_{t \to \infty} p_{ii}(t) = \lim_{\lambda \to 0} \lambda p_{ii}(\lambda) = (q_i \mathbb{E}_i \sigma_i)^{-1}.$$

b) Given $i \notin H$, we can choose $i_0, i_1, \cdots, i_n = i$ such that $i_0 \in H$ and

$$\bar{p}_{i_0 i_1} \cdots \bar{p}_{i_{n-1} i} > 0.$$

Since $e_{i_0 H}(0) < \infty$, the latter assertion follows from Lemmas 2.20 and 1.46.   □

Proof of Theorem 2.15   a) Assume the chain is ergodic and then recurrent. It follows from Lemma 2.17 that $f_{iH} = 1$ for every i. Then by Lemma 2.20, $(e_{iH}(\lambda) : i \in E)$ is the minimal solution of equation

$$x_i = \frac{q_i}{q_i - \lambda} \sum_{k \notin H} \bar{p}_{ik} x_k + \frac{1}{q_i - \lambda}, \qquad i \in E.$$

By the definition of ergodicity and Lemma 2.21, we have $e_{iH}(0) < \infty$ for each $i \in E$. Set $y_i = 0$ for $i \in H$ and $y_i = e_{iH}(0)$ for $i \notin H$. Then

$$\sum_j q_{ij} y_j = \sum_{j \notin H} q_{ij} e_{jH}(0) = -1$$

for $i \notin H$. At the same time

$$\sum_{i \in H} \sum_{j \neq i} q_{ij} y_j = \sum_{i \in H} \sum_{j \notin H} q_{ij} e_{jH}(0) = \sum_{i \in H} (q_i e_{iH}(0) - 1) < \infty.$$

Hence (2.12) holds.

b) The proof of the converse direction is similar to the discrete-time case. Let (x_i) be a solution of equation

$$\begin{cases} \sum_j q_{ij} x_j + 1 \leqslant 0, & i \notin H, \\[2mm] \sum_{i \in H} \sum_{j \neq i} q_{ij} x_j < \infty. \end{cases}$$

Define

$$c_i = \begin{cases} 1, & i \notin H; \\[2mm] -\sum_j q_{ij} x_j, & i \in H. \end{cases}$$

Then

$$c_i + \sum_j q_{ij} x_j \leqslant 0, \quad i \in E.$$

Thus for each $\lambda > 0$,

$$x_i - c \geqslant \sum_{j \neq i} \frac{q_{ij}}{\lambda + q_i} (x_j - c) + \frac{c_i - \lambda c}{\lambda + q_i}, \quad i \in E,$$

where $c = 0 \wedge \inf\{c_j / \lambda : j \in E\} > -\infty$.

c) Since $(p_{ij}(\lambda))$ satisfies the backward equation (2.6), from the comparison theorem (Theorem 1.42) it follows that

$$x_i - c \geqslant \sum_j p_{ij}(\lambda)(c_j - \lambda c), \quad i \in E.$$

Now, by regularity, $\lambda \sum_j P_{ij}(\lambda) = 1$, so that

$$x_i \geqslant \sum_j p_{ij}(\lambda) c_j, \quad i \in E, \ \lambda > 0.$$

Furthermore,

$$x_i \geqslant \sum_{j \notin H} p_{ij}(\lambda) + \sum_{j \in H} p_{ij}(\lambda) c_j = \lambda^{-1} + \sum_{j \in H} p_{ij}(\lambda)(c_j - 1).$$

That is,

$$\lambda x_i \geqslant 1 + \sum_{j \in H} \lambda p_{ij}(\lambda)(c_j - 1), \quad i \in E, \ \lambda > 0.$$

As

$$\lim_{\lambda \to 0} \lambda p_{ij}(\lambda) = \lim_{t \to \infty} p_{ij}(t) = \pi_j,$$

we see

$$0 \geqslant 1 + \sum_{j \in H} \pi_j(c_j - 1).$$

This shows that $\exists j \in H$ such that $\pi_j > 0$. Hence by irreducibility, $\pi_j > 0$ for every j. $\square$

2.3   Single Birth Processes and Birth-Death Processes

Till now, we have studied 4 basic topics on Q-processes:

(1) uniqueness,   (2) recurrence,   (3) ergodicity,   (4) exponential ergodicity.

All criteria depend upon test functions (y_i) or (ϕ_i), this is the best answer for general cases as one can expect. Those processes having criteria without using test functions are of two kinds: one is the single birth process, the other is the branching process.

Definition 2.22. $Q = (q_{ij} : i, j \in \mathbb{Z}_+)$ is called a **single birth** Q-**matrix**, if $q_{i,i+1} > 0$ and $q_{i,j} = 0$ for every $i \geqslant 0$ and $j - i \geqslant 2$. And it is called a **birth-death** Q-**matrix**, if $q_{i,i+1} =: b_i > 0 (i \geqslant 0)$, $q_{i,i-1} =: a_i > 0 (i \geqslant 1)$, and $q_{ij} = 0$ for every $|i - j| \geqslant 2$.

In the following, we will relax the condition by allowing

$$N := \max\{i + 1 : \ q_{i,i+1} = 0\} < \infty.$$

When $N = \max \varnothing = 0$, it is reduced to single birth Q-matrix and when $N \geqslant 1$, it is called the **single birth type** or **birth-death type** Q-matrix, respectively.

Uniqueness

Theorem 2.23. For every $n \geqslant N$, define

$$m_n = q_{n,n+1}^{-1} \left(1 + \sum_{j=0}^{N-1} q_{nj} + \sum_{k=N}^{n-1} m_k \sum_{j=0}^{k} q_{nj} \right),$$

and by convention $\sum\limits_{\varnothing} = 0$. Then the corresponding Q-process is unique iff $\sum\limits_{n=N}^{\infty} m_n = \infty$. In particular, for birth-death type, we have

$$m_n = \frac{1}{b_n} + \frac{a_N \cdots a_n}{b_N \cdots b_n} + \sum_{k=N}^{n-1} \frac{a_{k+1} \cdots a_n}{b_k \cdots b_n}, \quad n \geqslant N.$$

Proof a) We need only prove that for every $\lambda > 0$, the maximal solution (u_i^*) to equation

$$u_i = \sum_{j \neq i} q_{ij} u_j / (\lambda + q_i), \qquad 0 \leqslant u_i \leqslant 1, \quad i \geqslant 0$$

is null. When $N \geqslant 1$, the set $\{0, 1, \cdots, N-1\}$ forms a closed subset for the chain, so that for every $i \leqslant N-1$, $u_i^* = 0$.

b) Define $q_k^{(i)} = \sum\limits_{j=0}^{i} q_{kj}$ $(i < k, k \geqslant 1)$ and

$$\begin{cases} F_k^{(k)} = 1, & k \geqslant N; \\ F_k^{(i)} = \sum\limits_{j=i}^{k-1} q_k^{(j)} F_j^{(i)} \big/ q_{k,k+1}, & k > i \geqslant N. \end{cases} \tag{2.20}$$

Then

$$m_n = q_{n,n+1}^{-1} \big(1 + q_n^{(N-1)} + \sum_{k=N}^{n-1} q_n^{(k)} m_k \big), \qquad n \geqslant N.$$

By induction, we have

$$F_n^{(N)} \leqslant q_{N,N+1} m_n, \qquad n \geqslant N. \tag{2.21}$$

c) Suppose (u_i) satisfies equation:

$$(1 + q_i) u_i = \sum_{j \neq i} q_{ij} u_j, \quad i \geqslant 0; \quad u_k = 0 \text{ for } k \leqslant N-1 \text{ but } u_N = 1. \tag{2.22}$$

Applying a) to $\lambda = 1$, we need only prove that (u_i) is unbounded if and only if $\sum\limits_{n=N}^{\infty} m_n = \infty$. From (2.22), we see

$$u_{n+1} - u_n = q_{n,n+1}^{-1} \left[\sum_{k=0}^{n-1} q_n^{(k)} (u_{k+1} - u_k) + u_n \right], \qquad n \geqslant N. \tag{2.23}$$

Thus u_i is increasing in i.

The key of the proof is the following estimate:

$$m_k \leqslant u_{k+1} - u_k \leqslant (u_{N+1} - u_N) F_k^{(N)} + u_k m_k, \qquad k \geqslant N. \tag{2.24}$$

To prove (2.24), we use the induction method. Notice that

$$m_N = q_{N,N+1}^{-1}\big[1 + q_N^{(N-1)}\big] = u_{N+1} - u_N$$

and (2.23). Clearly,

$$u_{n+1} - u_n = q_{n,n+1}^{-1}\left[q_n^{(N-1)} + \sum_{k=N}^{n-1} q_n^{(k)}(u_{k+1} - u_k) + u_n\right], \qquad n \geqslant N.$$

Suppose (2.24) holds for k: $N \leqslant k \leqslant n - 1$. Now consider $k = n$. Then

$$u_{n+1} - u_n \geqslant q_{n,n+1}^{-1}\left[q_n^{(N-1)} + \sum_{k=N}^{n-1} q_n^{(k)} m_k + u_n\right] \geqslant m_n, \qquad n \geqslant N + 1,$$

and

$$u_{n+1} - u_n \leqslant q_{n,n+1}^{-1}\left[(u_{N+1} - u_N) \sum_{k=N}^{n-1} q_n^{(k)} F_k^{(N)} + q_n^{(N-1)}\right.$$

$$\left. + \sum_{k=N}^{n-1} q_n^{(k)} m_k u_k + u_n\right]$$

$$\leqslant (u_{N+1} - u_N) F_n^{(N)} + \frac{u_n}{q_{n,n+1}}\left[1 + q_n^{(N-1)} + \sum_{k=N}^{n-1} q_n^{(k)} m_k\right]$$

$$= (u_{N+1} - u_N) F_n^{(N)} + u_n m_n, \qquad n \geqslant N + 1.$$

d) Finally, we prove that (u_i) is bounded iff

$$R := \sum_{n=N}^{\infty} m_n < \infty.$$

Assume first that $u_\infty := \lim_{n \to \infty} u_n < \infty$. Then it follows from (2.24) that

$$R = \sum_{k=N}^{\infty} m_k \leqslant \lim_{n \to \infty} \sum_{k=N}^{n} (u_{k+1} - u_k) = u_\infty - 1 < \infty.$$

Conversely, suppose $R < \infty$. Since

$$u_\infty = \prod_{k=0}^{\infty} \frac{u_{k+1}}{u_k} \quad \text{and} \quad \frac{u_{k+1}}{u_k} \geqslant 1, \quad k \geqslant N,$$

we see that whenever $u_{k+1}/u_k - 1 \to 0$, the quantities

$$\prod_k \frac{u_{k+1}}{u_k} \quad \text{and} \quad \sum_k \log \frac{u_{k+1}}{u_k},$$

or

$$\sum_k \log \frac{u_{k+1}}{u_k} \quad \text{and} \quad \sum_k \left(\frac{u_{k+1}}{u_k} - 1\right)$$

diverge simultaneously. But from (2.24) and (2.21) it follows that

$$0 \leqslant u_{k+1} u_k^{-1} - 1 \leqslant (u_{N+1} - u_N) u_k^{-1} F_k^{(N)} + m_k$$
$$\leqslant [1 + (u_{N+1} - u_N) q_{N,N+1}] m_k.$$

Gathering these facts together, we can deduce the desired assertion. □

The next result shows that for uniqueness, the case of $N = 0$ is essential. The probabilistic proof is rather interesting.

Theorem 2.24. Assume $N \geqslant 1$. Choose an arbitrary positive sequence $(b_i : i \leqslant N - 1)$. If $q_{i,i+1} = 0$, define $\bar{q}_{i,i+1} = b_i$, $\bar{q}_i = q_i + b_i$; and for other $j \neq i$, define $\bar{q}_{ij} = q_{ij}$. Then (q_{ij})-process is unique iff so is $(\bar{q}_{ij})$-process. In other words, the case of $N \geqslant 1$ can be reduced to the case of $N = 0$.

Proof Denote by (X_t) and $(\bar{X}_t)$, respectively, the minimal processes corresponding to (q_{ij}) and $(\bar{q}_{ij})$. We need prove that during every finitely time interval, (X_t) jumps finite many times more than $(\bar{X}_t)$. Suppose $(\tilde{X}_t)$ is a minimal process, corresponding to Q-matrix $(\tilde{q}_{ij})$: when $i \leqslant N - 1$, $\tilde{q}_i = 0$; and for all $i \geqslant N$, $\tilde{q}_{ij} = q_{ij}$. Note that for every $i \leqslant N - 1$, $(\bar{X}_t)$ stays at i in exponential law with rate $\bar{q}_i$, then jumps to some other state. This is the only way that makes $(\bar{X}_t)$ produce more jumps than $(\tilde{X}_t)$. By conditional independence and the fact $N < \infty$, these jumps are only finite during any finite time interval. Therefore, during any finite time interval, $(\bar{X}_t)$ jumps at most finitely many times more than $(\tilde{X}_t)$. The same conclusion can also applied to (X_t) and $(\tilde{X}_t)$. Hence we obtain the desired assertion. □

Recurrence and Positive Recurrence. Hitting Probability and the First Moment

When $N = 0$, the following two theorems give the criteria for recurrence and positive recurrence, respectively. When $N \geqslant 1$, they give the criteria for hitting set $\{0, \cdots, N - 1\}$ with probability one and for finiteness of the first moment of the hitting time, respectively. For this, let

$$\tau_N = \inf\{t \geqslant 0 : X_t \leqslant N - 1\}.$$

Theorem 2.25. When $N = 0$, assume Q is irreducible. When $N \geqslant 1$, assume for every $i_0 \geqslant N$, there exist $i_1, \cdots, i_m$ such that $i_m \leqslant N - 1$, $q_{i_0 i_1} q_{i_1 i_2} \cdots q_{i_{m-1} i_m} > 0$. Then $\mathbb{P}^i[\tau_N < \infty] = 1$ for every $i \geqslant N$ (when $N \geqslant 1$) (respectively, $i \geqslant 1$ (when $N = 0$)) iff $\sum\limits_{n=N}^{\infty} F_n^{(N)} = \infty$, where $\left(F_n^{(k)}\right)$ is defined by (2.20). For the birth-death type,

$$F_N^{(N)} = 1, \quad F_n^{(N)} = \frac{a_{N+1} \cdots a_n}{(b_{N+1} \cdots b_n)}, \quad n > N.$$

Theorem 2.26. Under the assumptions of Theorem 2.25, for every $i \geqslant N$ (if $N \geqslant 1$) (respectively, $i \geqslant 1$ (if $N = 0$)), $\mathbb{E}^i \tau_N < \infty$ iff

$$d := \sup_{k \geqslant N} \sum_{s=N}^{k} d_s \Big/ \sum_{s=N}^{k} F_s^{(N)} < \infty,$$

where

$$d_N = 0, \quad d_n = \frac{1}{q_{n,n+1}} \left(1 + \sum_{s=N}^{n-1} d_s \sum_{j=0}^{s} q_{nj} \right), \qquad n > N.$$

In particular, for the birth-death type, $d_N = 0$ and

$$d_n = 1/b_n + \sum_{k=N+1}^{n-1} a_{k+1} \cdots a_n / (b_k \cdots b_n), n > N.$$

Proofs of Theorems 2.25 and 2.26

a) We first prove that case $N \geqslant 1$ can be reduced into case $N = 0$.

Without loss of generality, we can view $\{0, \cdots, N-1\}$ as a singleton $\{0\}$. The resulted Markov chain is a single birth process with absorbing state 0. Thus, we can reduce case $N \geqslant 2$ to case $N = 1$.

Given a Markov chain, we change a state, say 0, to an absorbing state, and obtain a new Markov chain $(\widetilde{X}_t)$. Set $\tilde{\tau}_0 = \inf\{t \geqslant 0 : \widetilde{X}_t = 0\}$. Then by Theorem 2.13, Proposition 1.46 and Theorem 2.15, the original Markov chain is recurrent (respectively, positive recurrent) iff $\mathbb{P}^i[\tilde{\tau}_0 < \infty] = 1$ (respectively, $\mathbb{E}^i \tilde{\tau}_0 < \infty$). Therefore, we reduce case $N = 1$ to case $N = 0$.

b) Prove that if $Q = (q_{ij})$ is a regular and irreducible Q-matrix, then the corresponding $P(t)$ is recurrent iff for some (equivalently, all) j_0, equation

$$x_i = \sum_{k \neq j_0} q_{ik} x_k / q_i, \quad 0 \leqslant x_i \leqslant 1, \quad i \in E \tag{2.25}$$

has only null solution.

Indeed, by Theorem 2.13, Lemma 1.46 and Lemma 2.17, we need only prove the minimal solution (x_i^*) to equation

$$x_i = \sum_{k \neq j_0} q_{ik} x_k / q_i + q_{ij_0} / q_i, \quad i \in E \tag{2.26}$$

equals 1. Note that equation (2.25) is the homogeneous part of equation (2.26). On the other hand, since $(x_i = 1 : i \in E)$ is the nonnegative solution to equation (2.26), for all i, we have $x_i^* \leqslant 1$. Gathering the facts together, we obtain the desired assertion.

c) To prove the recurrence, we need only prove that whenever $\sum\limits_{k=0}^{\infty} F_k^{(0)} <$ ∞, equation (2.25) has nontrivial solution. Choose $j_0 = 0$. Then equation (2.25) has nontrivial solution iff equation

$$x_i = \sum_{k \neq 0} q_{ik} x_k / q_i, \quad x_0 = 1, \ i \in E$$

has nonnegative bounded solution. However the latter equation has unique solution:

$$x_0 = 1 = x_1,$$

$$x_{i+1} - x_i = \left(\sum_{j=1}^{i-1} q_{ij}(x_i - x_j) + q_{i0}x_i \right) \Big/ q_{i,i+1}$$

$$= \left(\sum_{j=0}^{i-1} q_{ij}(x_i - x_j) + q_{i0} \right) \Big/ q_{i,i+1}, \quad i \geqslant 1.$$

Clearly, $\{x_i\}$ is increasing in i. Thus, the proof is reduced to show that (x_i) is bounded iff $\sum\limits_{k=0}^{\infty} F_k^{(0)} < \infty$. Let $U_n = \sum\limits_{k=1}^{n} u_k$. By using integration by parts formula,

$$\sum_{k=1}^{n} u_k v_k = \sum_{k=1}^{n} (U_k - U_{k-1}) v_k = \sum_{k=1}^{n-1} U_k (v_k - v_{k+1}) + U_n v_n,$$

we have

$$\sum_{j=0}^{i-1} q_{ij}(x_i - x_j) = \sum_{j=0}^{i-1} q_i^{(j)}(x_{j+1} - x_j) = \sum_{j=1}^{i-1} q_i^{(j)}(x_{j+1} - x_j).$$

Thus

$$x_{i+1} - x_i = \sum_{j=1}^{i-1} q_i^{(j)}(x_{j+1} - x_j)/q_{i,i+1} + q_i^{(0)}/q_{i,i+1}, \quad i \geqslant 1.$$

This shows that $(y_i := x_{i+1} - x_i)_{i \geqslant 1}$ is a nonnegative solution to equation

$$y_i = \sum_{j=1}^{i-1} q_i^{(j)} y_j / q_{i,i+1} + q_i^{(0)}/q_{i,i+1}, \quad i \geqslant 1.$$

But this equation has unique solution:

$$z_1 = q_1^{(0)}/q_{12}, \quad z_i = \sum_{j=1}^{i-1} q_i^{(j)} z_j / q_{i,i+1} + q_i^{(0)}/q_{i,i+1}, \quad i \geqslant 2.$$

By induction, we easily obtain

$$z_i = F_i^{(0)}, \quad i \geqslant 1.$$

Gathering the facts above, we deduce

$$x_0 = x_1 = 1 = F_0^{(0)}, \quad x_{i+1} - x_i = F_i^{(0)}, \quad i \geqslant 1,$$

which implies the recurrence.

d) Finally we prove positive recurrence. For this, we apply Theorem 2.15. Let (u_i) $(u_0 = 0)$ be a nonnegative solution to equation (2.12) (by setting $H = \{0\}$). Then

$$q_{k,k+1}(u_{k+1} - u_k) + 1 = \sum_j q_{kj}u_j + 1 - \sum_{j=0}^{k-1} q_{kj}u_j - q_{k,k+1}u_k + q_k u_k$$

$$\leqslant \sum_{j=0}^{k-1} q_{kj}(u_k - u_j) = \sum_{j=0}^{k-1} q_k^{(j)}(u_{j+1} - u_j).$$

From this, by induction we have

$$v_k \leqslant F_k^{(0)}v_0 - d_k, \quad k \geqslant 0,$$

where

$$v_k = u_{k+1} - u_k, \quad k \geqslant 0.$$

Thus

$$0 \leqslant u_{k+1} = \sum_{s=0}^{k} v_s \leqslant \left(\sum_{s=0}^{k} F_s^{(0)}\right)v_0 - \sum_{s=0}^{k} d_s = \left(\sum_{s=0}^{k} F_s^{(0)}\right)u_1 - \sum_{s=0}^{k} d_s.$$

This gives $d \leqslant u_1 < \infty$. Conversely, suppose $d < \infty$. Let

$$u_0 = 0, \quad u_1 \in [d, \infty), \quad u_{k+1} = u_k + F_k^{(0)}u_1 - d_k, \quad k \geqslant 1.$$

Obviously,

$$\sum_{j \neq 0} q_{0j}u_j = q_{01}u_1 < \infty.$$

And for every $k > 0$,

$$1 + q_{k,k+1}(u_{k+1} - u_k) = q_{k,k+1}F_k^{(0)}u_1 - q_{k,k+1}d_k + 1$$

$$= \sum_{s=0}^{k-1} q_k^{(s)}F_s^{(0)}u_1 - \sum_{s=0}^{k-1} q_k^{(s)}d_s = \sum_{s=0}^{k-1} q_k^{(s)}\left(F_s^{(0)}u_1 - d_s\right)$$

$$= \sum_{s=0}^{k-1} q_k^{(s)}(u_{s+1} - u_s) = \sum_{j=0}^{k-1} q_{kj}(u_k - u_j).$$

Hence, $\sum_j q_{kj} u_j + 1 = 0$ for $k > 0$.

e) To prove the last assertion for the birth-death process, note that $m_n = F_n^{(0)}/b_0 + d_n$. Then

$$\sum_{s=0}^{k} d_s \bigg/ \sum_{s=0}^{k} F_k^{(0)} = \sum_{s=0}^{k} m_s \bigg/ \sum_{s=0}^{k} F_s^{(0)} - 1/b_0.$$

We see that if the process is unique and the conditions in Theorem 2.26 are fulfilled, then we have $\sum_{k=0}^{\infty} F_k^{(0)} = \infty$. Thus Stoltz' theorem implies that

$$\lim_{k\to\infty} \frac{\sum_{s=0}^{k} d_s}{\sum_{s=0}^{k} F_s^{(0)}} = \lim_{k\to\infty} \frac{d_k}{F_k^{(0)}} = \frac{1}{b_0} \sum_{k=0}^{\infty} \frac{b_0 \cdots b_k}{a_1 \cdots a_{k+1}}.$$

Conversely, if the right-hand side in the previous equation converges, then the conditions of the theorem are fulfilled by using the property of proportional ratio.   □

Next, we present a simple sufficient condition that $\mathbb{E}^i \tau_N < \infty$ for every $i \geqslant N$.

Theorem 2.27. Assume that the conditions in Theorem 2.26 are fulfilled. If there exist constants $c_1 \geqslant c_2 \geqslant 0$ such that

$$M_n := c_1 \sum_{j=0}^{N} q_{nj} + c_2 \left[\sum_{k=N+1}^{n-1} \sum_{j=0}^{k} q_{nj} - q_{n,n+1} \right] \geqslant 1, \qquad n \geqslant N+1, \quad (2.27)$$

then $\mathbb{E}^i \tau_N < \infty$.

Proof Define $G_n = c_1 F_n^{(N)} - c_2$, $n \geqslant N$. By (2.20), we have

$$G_n = c_1 \sum_{j=N}^{n-1} q_n^{(j)} F_j^{(N)} \big/ q_{n,n+1} - c_2$$

$$= c_1 q_n^{(N)} \big/ q_{n,n+1} + c_1 \sum_{j=N+1}^{n-1} q_n^{(j)} F_j^{(N)} \big/ q_{n,n+1} - c_2$$

$$= c_1 q_n^{(N)} \big/ q_{n,n+1} + \sum_{j=N+1}^{n-1} q_n^{(j)} G_j \big/ q_{n,n+1} + c_2 \sum_{j=N+1}^{n-1} q_n^{(j)} \big/ q_{n,n+1} - c_2$$

$$= M_n \big/ q_{n,n+1} + \sum_{j=N+1}^{n-1} q_n^{(j)} G_j \big/ q_{n,n+1}.$$

From this and the definition of d_n (see Theorem 2.25), it follows that

$$\begin{cases} G_n = M_n/q_{n,n+1} + \displaystyle\sum_{j=N+1}^{n-1} q_n^{(j)} G_j/q_{n,n+1}, \\ d_n = 1/q_{n,n+1} + \displaystyle\sum_{j=N+1}^{n-1} q_n^{(j)} d_j/q_{n,n+1}, \qquad n \geqslant N+1. \end{cases} \tag{2.28}$$

Since

$$G_N = c_1 - c_2 \geqslant 0 = d_N$$

and by (2.27),

$$G_{N+1} = M_{N+1}/q_{N+1,N+2} \geqslant 1/q_{N+1,N+2} = d_{N+1}.$$

Using (2.27) and (2.28) again, by induction, we have

$$d_n \leqslant G_n = c_1 F_n^{(N)} - c_2 \leqslant c_1 F_n^{(N)}, \quad n \geqslant N.$$

Therefore,

$$\sup_{n \geqslant N} \sum_{k=N}^{n} d_k \bigg/ \sum_{k=N}^{n} F_k^{(N)} \leqslant \sup_{n \geqslant N} d_n \bigg/ F_n^{(N)} \leqslant c_1 < \infty. \quad \square$$

2.4   Branching Processes and Extended Branching Processes

Branching Processes

A classical branching process is a continuous-time Markov chain with the following Q-matrix:

$$q_{ij} = \begin{cases} i\lambda p_0, & j = i-1; \\ -i\lambda(1-p_1), & j = i; \\ i\lambda p_{j-i+1}, & j \geqslant i+1; \\ 0, & j < i-1, \end{cases}$$

where $(p_i, i \geqslant 0)$ is a probability distribution. The background for this branching process is that assuming at time 0 there are X_0 particles, the splitting times of every particle are independent, identically distributed as exponential law with parameter λ. And the splitting particles ("next generation") from each particle are also independent, with the distribution sequence $(p_i, i \geqslant 0)$. Let X_t be the total particles at time t.

Alike the discrete-time branching process, 0 is an absorbing state (as $q_0 = 0$). What we concern is the so-called extinct probability:

$$r_i = \lim_{t\to\infty} p_{i0}(t).$$

By independence, we have $r_i = r_1^i$, so that we need only consider $r := r_1$.
Let

$$\phi(t,s) = \sum_{j=0}^{\infty} p_{1j}(t)s^j, \quad t \geqslant 0, s \in (0,1),$$

then it is easy to see

$$\sum_{j=0}^{\infty} p_{ij}(t)s^j = \phi(t,s)^i, \quad \phi(t+u,s) = \phi(t,\phi(u,s)).$$

Fix $h > 0$. Consider the h-skeleton process $Y_n = X_{nh}, n \geqslant 0$. Then

$$\mathbb{P}[\exists n, Y_n = 0] = \lim_{n\to\infty} \mathbb{P}[Y_n = 0] = \lim_{n\to\infty} \mathbb{P}[X_{nh} = 0] = \lim_{t\to\infty} \mathbb{P}[X_t = 0] = r.$$

Thus by Theorem 1.52, we know that for every $t > 0$, r is the minimal
nonnegative solution to equation $\phi(t,s) = s$.

Since $p_{ij}(t) = \delta_{ij} + q_{ij}t + o(t)$ as $t \to 0$ and $\phi(0,r) = r$, we have

$$\phi(t,s) = \sum_{j=0}^{\infty} [\delta_{1j} + q_{1j}t + o(t)]s^j = s + \lambda t[P(s) - s] + o(t).$$

By letting $t \to 0$, we obtain the following theorem.

Theorem 2.28. Denote $P(s) = \sum\limits_{k=0}^{\infty} p_k s^k$. Then r is the minimal nonnegative

solution for equation $P(s) = s, s \in [0,1]$. Therefore, $r = 1$ iff $\sum\limits_{k=1}^{\infty} kp_k \leqslant 1$.

Extended Branching Process

In many situations, the independence is not fulfilled, that is, the particles may interact each other. So we shall consider more general branching
processes. Suppose Q-matrix has the following form:

$$q_{ij} = \begin{cases} a_j, & i = 0; \\ i^\nu \lambda p_0, & i \geqslant 1, j = i-1; \\ -i^\nu \lambda(1-p_1), & i \geqslant 1, j = i; \\ i^\nu \lambda p_{j-i+1}, & i \geqslant 1, j \geqslant i+1; \\ 0, & \text{for other } i, j, \end{cases} \tag{2.29}$$

where $\nu > 0$, $\lambda > 0$, $a_0 < 0$, $\sum\limits_{i=0}^{\infty} a_i = 0$; and the probability distribution
$(p_i : i \geqslant 0)$ is the same as above. When $\nu = 1$, it is called linear; when
$\nu < 1$, it is called sublinear; when $\nu > 1$, it is called super-linear. Without
loss of generality, assume $\lambda = 1$.

For uniqueness, we have

Theorem 2.29. If $\rho := \sum\limits_{k=1}^{\infty} k p_k \leqslant 1$, then the extended branching process is unique; if $\nu > 1$, then condition $P'(1) \leqslant 1$ is also necessary for the uniqueness.

Proof Prove first when $\rho \leqslant 1$, the process is unique. By Theorem 2.24, we can assume $q_0 = 0$. By taking $E_n = \{1, 2, \cdots, n\}$ and $\phi_i = i + 1$ in Theorem 2.9, we have

$$\sum_j q_{ij}(\phi_j - \phi_i) = -q_{ii-1} + \sum_{j=i+1}^{\infty} q_{ij}(j - i) = \lambda i^{\nu}(\rho - 1) \leqslant 0, \quad i \geqslant 1.$$

Thus $\sum\limits_j q_{ij}(\phi_j - \phi_i) \leqslant \phi_i$ for each $i \geqslant 0$.

Assume $\nu > 1$ and $\rho > 1$. To prove the process is not unique, we compare it with a birth-death process. Let $\Gamma \in (p_0, p_0 + \rho - 1)$ to be determined later. Construct a birth-death process with birth rates $b_0 = 0, b_i = \gamma_i \Gamma (i \geqslant 1)$ and death rates $a_i = \gamma_i p_0 (i \geqslant 1)$. Since

$$\sum_{n=1}^{\infty} \left(\frac{1}{b_n} + \sum_{k=1}^{n-1} \frac{a_{k+1} \cdots a_n}{b_k \cdots b_n} \right) = \frac{1}{\Gamma} \sum_{n=1}^{\infty} \frac{1}{\gamma_n} \sum_{n=0}^{\infty} \left(\frac{p_0}{\Gamma} \right)^n < \infty,$$

by Theorem 2.23, we see that this birth-death process is not unique. Thus equation

$$a_i(u_{i-1} - u_i) + b_i(u_{i+1} - u_i) = \lambda u_i, \quad u_0 = 0, \lambda > 0, i \geqslant 1$$

has nonnegative (non-null) bounded solution. It is obviously to see that $u_i \uparrow$, and

$$u_{i=1} - u_i \geqslant \frac{p_0}{\Gamma}(u_i - u_{i-1}), \quad i \geqslant 1. \tag{2.30}$$

By letting $\Gamma \downarrow p_0$, we have

$$\sum_{k=1}^{\infty} p_{k+1} \sum_{\ell=0}^{k-1} \left(\frac{p_0}{\Gamma} \right)^{\ell} \uparrow \sum_{k=1}^{\infty} k p_{k+1} = p_0 + \rho - 1 > p_0.$$

Thus we can choose $\Gamma \in (p_0, p_0 + \rho - 1)$ such that

$$\sum_{k=1}^{\infty} p_{k+1} \sum_{\ell=0}^{k-1} \left(\frac{p_0}{\Gamma} \right)^{\ell} \geqslant \Gamma. \tag{2.31}$$

From (2.30) and (2.31), we have for $i \geqslant 1$,

$$\sum_j q_{ij}(u_j - u_i) = a_i(u_{i-1} - u_i) + \gamma_i \sum_{k=1}^{\infty} \sum_{\ell=1}^{k}(u_{i+\ell} - u_{i+\ell-1})$$

$$\geqslant a_i(u_{i-1} - u_i) + \gamma_i(u_{i+1} - u_i) \sum_{k=1}^{\infty} p_{k+1} \sum_{\ell=0}^{k-1} \left(\frac{p_0}{\Gamma} \right)^{\ell}$$

$$\geqslant a_i(u_{i-1} - u_i) + b_i(u_{i+1} - u_i) = \lambda u_i.$$

By the comparison theorem (Theorem 1.42) and the uniqueness theorem (Theorem 2.7), we see the original process is not unique. $\square$

For recurrence, we have

Theorem 2.30. The extended branching process is recurrent iff $\rho \leqslant 1$.

Proof By Theorem 2.13, we need only consider the recurrence of its embedded chain $\bar{P} = (\bar{p}_{ij})$. Note

$$\bar{p}_{ij} = \begin{cases} p_{j-i+1}(1-p_1)^{-1}, & j \neq i,\ j \geqslant i-1 \geqslant 0; \\ 0, & \text{else.} \end{cases}$$

is independent of ν. Thus we know the assertion holds from the necessary and sufficient condition proven in the previous section. $\square$

For the ergodicity, we have

Theorem 2.31. Let $A(s) = \sum\limits_{j=0}^{\infty} a_j s^j$, for $s \in [0,1]$. Assume $\nu \geqslant 1$ and $\rho \leqslant 1$. Then the extended branching process is ergodic iff

$$\int_0^1 \frac{A(s)}{P(s)-s}(1-s)^{\nu-1} ds > -\infty. \tag{2.32}$$

Proof To prove the ergodicity, we will consider the following equation:

$$\mu_j q_j = \sum_{i \neq j} \mu_i q_{ij}, \quad \mu_j > 0,\ j \geqslant 0,\ \sum_j \mu_j < \infty. \tag{2.33}$$

a) First, we prove that the solution (μ_i) to this equation is decreasing so that (μ_i) is bounded. Indeed, since

$$1 \geqslant \rho = \sum_{k \geqslant 0} k p_k = \sum_{k \geqslant 0}(k-1)p_k + 1, \text{and} \quad p_0 \geqslant \sum_{k \geqslant 2} p_k,$$

so that $2^\nu p_0 + p_1 \geqslant 1$ and

$$\mu_2 = \frac{\mu_1 q_1 - \mu_0 q_{01}}{q_{21}} \leqslant \frac{\mu_1 q_1}{q_{21}} = \mu_1 \frac{1-p_1}{2^\nu p_0} \leqslant \mu_1.$$

Suppose $\mu_j \leqslant \mu_{j-1} \leqslant \cdots \leqslant \mu_1$. Then for $j \geqslant 2$ we have

$$q_{j+1,j}\mu_{j+1} = \mu_j q_j - \sum_{i=0}^{j-1} \mu_i q_{ij} \leqslant \mu_j \left(q_j - \sum_{i=1}^{j-1} q_{ij} \right).$$

Note by the mean value theorem,

$$q_{j+1,j} - q_j + \sum_{i=1}^{j-1} q_{ij} = p_0[(j+1)^\nu - j^\nu] - \sum_{k=2}^{j}[j^\nu - (j+1-k)^\nu]p_k - j^\nu \sum_{k=j+1}^{\infty} p_k$$

$$\geqslant \nu j^{\nu-1} p_0 - \nu j^{\nu-1} \sum_{k=2}^{j}(k-1)p_k - \nu j^{\nu-1} \sum_{k=j+1}^{\infty}(k-1)p_k$$

$$= \nu j^{\nu-1}(1-M) \geqslant 0.$$

Thus $\mu_{j+1} \leqslant \mu_j$.

b) Since

$$\mu_j q_j = \sum_{i=0}^{j-1} \mu_i q_{ij} + \mu_{j+1} q_{j+1,j},$$

we have

$$\mu_j j^\nu (1 - p_1) = \mu_0 a_j + \sum_{i=1}^{j-1} \mu_i i^\nu p_{j-i+1} + \mu_{j+1}(j+1)^\nu p_0, \quad j \geqslant 1. \quad (2.34)$$

Firstly multiply both sides of (2.34) with s^j, and make a summation over $j \geqslant 1$ to derive

$$\sum_{j=1}^\infty \mu_j j^\nu s^{j-1} = \frac{\mu_0 A(s)}{P(s) - s}. \quad (2.35)$$

Then multiply both sides of (2.35) with $(1 - s)^{\nu-1}$, and integrate with respect to $s \in (0, 1)$ to derive

$$\sum_{j=1}^\infty \mu_j j^\nu \int_0^1 s^{j-1} (1-s)^{\nu-1} \mathrm{d}s = \int_0^1 \frac{\mu_0 A(s)}{P(s) - s} (1-s)^{\nu-1} \mathrm{d}s.$$

Note

$$\int_0^1 s^{j-1} (1-s)^{\nu-1} \mathrm{d}s = \frac{\Gamma(j)\Gamma(\nu)}{\Gamma(j+\nu)} \sim j^\nu \quad \text{as } j \to \infty.$$

Hence $\sum_j \mu_j < \infty$ iff

$$\int_0^1 \frac{A(s)}{P(s) - s} (1-s)^{\nu-1} \mathrm{d}s < \infty.$$

In this case, $\pi_i = \mu_i \bigg/ \sum_{i \geqslant 0} \mu_i$ is the stationary distribution.   □

2.5   Supplements and Exercises

(1) (Poisson process) The transition probability matrix $P(t) = (p_{ij}(t))$ of Poisson process $(X_t)_{t \geqslant 0}$ is:

$$p_{ij}(t) = \begin{cases} \frac{(\lambda t)^{j-i}}{(j-i)!} e^{-\lambda t}, & \text{if } j - i \geqslant 0; \\ 0, & \text{else.} \end{cases}$$

Find its Q-matrix $Q = (q_{ij})$, and prove directly that

$$P(t) = e^{Qt} = \sum_{n=0}^\infty \frac{(Qt)^n}{n!}.$$

(2) (Pure birth process) The Q-matrix of the pure birth process $(X_t)_{t \geqslant 0}$ is as follows: $q_i = q_{ii+1} = b_i > 0$ $(i \geqslant 0)$, $q_{ij} = 0$ $(i \neq j)$. Let T_n the nth jump epoch, and $T = \lim_{n \to \infty} T_n$. Then $\mathbb{P}_0[T = \infty] = 1$ iff $\sum_n b_n^{-1} = \infty$.

   Hint: Consider $\mathbb{E}e^{-T}$.

(3) For a Markov chain on finite state space, both forward equation and backward equation hold, and

$$P(t) = e^{Qt} = \sum_{n=0}^{\infty} \frac{(Qt)^n}{n!}.$$

(4) For an irreducible and honest Markov chain $(p_{ij}(t))$ on finite state space, $\lim_{t \to \infty} p_{ij}(t) = \pi_j > 0$ and $\sum_j \pi_j = 1$.

(5) Prove Theorem 2.5.

   Hint: By $p_{ij}(s+t) = \sum_k p_{ik}(s)p_{kj}(t)$ and Fatou's lemma, it is easy to see $p_{ij}'(t) \geqslant \sum_k q_{ik}p_{kj}(t)$. On the other hand, for $N > i$,

$$\sum_{k \neq i} p_{ik}(s)p_{kj}(t) \leqslant \sum_{k \leqslant N, k \neq i} p_{ik}(s)p_{kj}(t) + \sum_{k > N} p_{ik}(s)$$

$$= \sum_{k \leqslant N, k \neq i} p_{ik}(s)p_{kj}(t) + 1 - p_{ii}(s) - \sum_{k \leqslant N, k \neq i} p_{ik}(s).$$

   Divide both sides by s, and then take limit $\varlimsup_{s \downarrow 0}$.

(6) Consider an irreducible continuous-time Markov chain on finite state space, its Q-matrix is $q_{ij}, i, j = 1, 2, \cdots, N$ satisfies $q_{ij} = q_{ji}$. Given any initial state, let $P_i(t) = \mathbb{P}[X_t = i]$ be the probability of i as time t. Define

$$E(t) = -\sum_{i=1}^{N} P_i(t) \log P_i(t).$$

   Prove that $E(t)$ is a non-decreasing function in $t \geqslant 0$.

(7) For an irreducible continuous-time Markov chain with conservative Q-matrix of cardinality $N < \infty$, prove that the rank of its Q-matrix is $N - 1$.

(8) Consider an honest continuous-time Markov chain on two-point state space $\{0, 1\}$. The sojourn time in 0 is distributed exponentially with parameter λ, while the sojourn time in 1 is distributed exponentially with parameter μ. Find $p_{00}(t)$.

(9) Prove directly that (2.2) $\Leftrightarrow$ (2.4) and (2.3) $\Leftrightarrow$ (2.5).

(10) Assume that Q-matrix of $P(t)$ is bounded, that is, $q_i \leqslant M < \infty$ for every i. Then $P(t)$ satisfies the forward equation.

(11) If Q is bounded, then when $t \downarrow 0$, $P(t)$ converges uniformly to I, that is $\lim_{t \downarrow 0} \sup_{ij} |p_{ij}(t) - \delta_{ij}| = 0$.

Hint: For matrix $A = (a_{ij})$, define $||A|| = \sup_{ij} |a_{ij}|$, then $||P_t - I|| \leqslant e^{t||Q||} - 1$.

(12) Let $p_{ij}(\lambda) = \int_0^\infty e^{-\lambda t} p_{ij}(t) dt$. Then

(a) $\forall t \geqslant 0, p_{ij}(t) \geqslant 0 \Rightarrow \forall \lambda > 0, p_{ij}(\lambda) \geqslant 0$;

(b) $\forall t \geqslant 0, \sum_j p_{ij}(t) \leqslant 1 \Rightarrow \forall \lambda > 0, \lambda \sum_j p_{ij}(\lambda) \leqslant 1$;

(c) $\forall t, s \geqslant 0, P(t + s) = P(t)P(s) \Rightarrow \forall \lambda, \mu > 0, P(\lambda) - P(\mu) + (\lambda - \mu)P(\lambda)P(\mu) = 0$;

(d) $\lim_{t \to 0} p_{ij}(t) = \delta_{ij} \Rightarrow \lim_{\lambda \to \infty} (\lambda p_{ij}(\lambda) - \delta_{ij}) = 0$;

(e) $P'(t)|_{t=0} = Q \Rightarrow \lim_{\lambda \to \infty} \lambda(\lambda p_{ij}(\lambda) - \delta_{ij}) = q_{ij}$.

(13) Prove that equations (2.6) and (2.7) have the same minimal nonnegative solution.

(14) Assume Q is totally stable and conservative. The Q-process is unique iff its minimal process is honest.

Hint: Note that $\lambda \sum_{j \in E} p_{ij}(\lambda)$ is the minimal solution of

$$x_i = \sum_{k \neq i} \frac{q_{ik}}{\lambda + q_i} + \frac{\lambda}{\lambda + q_i}.$$

(15) By using Markov property, prove

(a) $p_{ii}(t)$ is always positive on $[0, \infty)$ for every i;

(b) If there exists $T > 0$ such that $p_{ij}(T) > 0$, then $p_{ij}(t) > 0$ for every $t \geqslant T$.

(16) (Renewal process) Assume that $(X_n > 0)_{n \geqslant 1}$ are independent identically distributed (iid) random variables. Let

$$S_n = X_1 + X_2 + \cdots + X_n, \quad n \geqslant 1, \quad \text{by convention } S_0 = 0.$$

Define $N_t = \#\{n : 0 < S_n \leqslant t\}$, then $(N_t)_{t \geqslant 0}$ is called a renewal process.

(a) If X_1 is exponentially distributed with parameter λ, then N_t is Poisson process with parameter λ. Prove $\mathbb{E}N_t = \text{Var}N_t = \lambda t$.

(b) If $M_t := \mathbb{E}N_t, F_n(x) = \mathbb{P}[S_n \leqslant x]$, then

$$M_t = \sum_{k=1}^\infty F_k(t).$$

(17) (Queueing theory) Suppose the customers wait for their services (for example, wait to pay out in supermarket). Denote by X_t the length of queue at time t (i.e. number of customers). The periods between any two successive customer are independent and identically distributed, which are exponential distribution with parameter λ, and the service times are independent and identically distributed, which are exponential distribution with parameter μ. Try to study the recurrence and ergodicity for the process.

(18) Let $\tau_1, \tau_2, \cdots$ be the successive jump epoch for a continuous-time Markov chain, that is, $\tau_1 = \inf\{t \geqslant 0 : X_t \neq X_0\}$ be the epoch of the first jump, $\tau_n = \inf\{t > \tau_{n-1} : X_t \neq X_{t-}\}$ be the epoch of the nth jump. Then

(a) Let $\tau_0 = 0, \Delta\tau_n = \tau_n - \tau_{n-1}$. Then $\{\Delta\tau_n\}$ are iid;

(b) If define $Y_n = X_{\tau_n}$, then (Y_n) is a discrete-time Markov chain with transition probability matrix $\bar{P} = (\bar{p}_{ij})$ given below Definition 2.12.

(19) Prove that if the discrete-time Markov chain with transition probability matrix $P(h)$ is recurrent for each $h > 0$ iff $\int_0^\infty p_{ii}(t)dt = \infty$.
Hint: By Exercise 15, we have $\delta(h) = \min_{0 \leqslant t \leqslant h} p_{ii}(t) > 0$, so by Markov property,

$$\min_{nh \leqslant t \leqslant (n+1)h} p_{ii}(t) \geqslant p_{ii}(nh)\delta(h),$$

$$\max_{nh \leqslant t \leqslant (n+1)h} p_{ii}(t) \leqslant p_{ii}((n+1)h)/\delta(h).$$

(20) From $C - K$ equation it follows that for every $i, h \geqslant 0, t \geqslant 0$,

$$\sum_j |p_{ij}(t+h) - p_{ij}(t)| \leqslant 2(1 - p_{ii}(h)),$$

so that $p_{ij}(t)$ is uniformly continuous on $t \in [0, \infty)$ for every i, j. From this, prove that the discrete-time $P(h)$ is ergodic for each $h > 0$ iff $\lim_{t \to \infty} p_{jj}(t) = \pi_j > 0$.

(21) For $A \subset E$, let $\tau_A = \inf\{t \geqslant 0 : X_t \in A\}$ be the hitting time of A.

(a) Let $x_i = \mathbb{P}_i[\tau_A < \infty]$. Then $\sum_k q_{ik}x_k = 0$ for $i \notin A$.

(b) Assume the chain is recurrent. Let $y_i = \mathbb{E}_i\tau_A$. Then $(y_i : i \in E)$ is the minimal nonnegative solution to equation:

$$y_i = 0, \ i \in A; \quad \sum_{k \in E} q_{ik}y_k = -1, \ i \notin A.$$

(22) For Markov chain $(X_t)_{t \geq 0}$, fix $i \in E$ and let $T = |\{t \geq 0 : X_t = i\}|$, where $|A|$ is the length of the set A.

   (a) If i is a recurrent state, then $\mathbb{P}_i[T = +\infty] = 1$.

   (b) If i is a transient state, then $\mathbb{P}_i[T < +\infty] = 1$, and T is exponentially distributed.

(23) (Continuation of Exercise 18) Assume $(X_t)_{t \geq 0}$ has stationary distribution α, and its embedded chain $(Y_n)_{n \geq 0}$ has stationary distribution π. Try to find the relationship between α and π.

(24) (Linear population growth model) Consider the following birth-death process: $a_n = \mu n, b_n = \lambda n + a, \lambda, \mu, a > 0$. Prove that the expectation of population satisfies differential equation:

$$M_t' = a + (\lambda - \mu)M_t.$$

Thus $\lim\limits_{t \to \infty} M_t = \infty$ for $\lambda \geq \mu$, and $\lim\limits_{t \to \infty} M_t = \dfrac{a}{\mu - \lambda}$ for $\lambda < \mu$.

(25) $P(t)$ is irreducible iff so does Q.

(26) If Markov chain is positive recurrent, then (2.12) holds.

(27) For a regular and irreducible Markov chain, prove that

   (a) if $\sup\limits_{i} q_i < \infty$, and the jump process is ergodic, then its embedded chain is ergodic;

   (b) if $\inf\limits_{i} q_i > 0$, and the embedded chain of the jump process is ergodic, then the jump process is ergodic.

(28) Consider a birth-death process with $a_i = b_i, i \geq 1$, and $b_i > 0$ such that $\sum\limits_{i} 1/b_i < \infty$. Prove the process is ergodic, but its embedded chain is not ergodic.

(29) Assume $p_i > 0, \sum\limits_{i=.}^{\infty} p_i = 1,$. Let $\bar{p}_{0i} = p_i, i \geq 0; \bar{p}_{i0} = 1, i \geq 1, \bar{p}_{ij} = 0$ for other i, j. Prove the discrete-time Markov chain $\bar{P} = (\bar{p}_{ij})$ is positive recurrent. If let $q_i = p_i, q_{ij} = q_i \bar{p}_{ij} \ (i \neq j)$, then this Q-process is not positive recurrent.

(30) Consider a conservative Q-matrix such that $q_{0k} = b_k > 0$, $q_k = q_{k,k-1} > 0$ for $k \geq 1$, $q_0 = \sum\limits_{k \geq 1} q_{0k}$; and $q_{ij} = 0$ for other $j \neq i$.

   (a) Try to study the uniqueness and recurrence for Q-process.

   (b) If

$$0 < c_1 = \inf_{i \geq 1} q_i \leq \sup_{i \geq 1} q_i = c_2 < \infty,$$

then the process is ergodic iff $\sum\limits_{k \geq 1} k q_{0k} < \infty$.

(31) Consider a conservative Q-matrix such that $q_{0k} = b_k > 0, q_k = q_{k,0} > 0$ for $k \geqslant 0$, $q_{ij} = 0$ for ether $j \neq i$. Try to study the uniqueness, recurrence and ergodicity of Q-process.

(32) For a signed measure μ on E, define its total variance

$$||\mu||_{\mathrm{Var}} = \sup_{|f| \leqslant 1} \left| \sum_i \mu_i f_i \right|.$$

Prove

(a) $||\mu||_{\mathrm{Var}} = \sum_i |\mu_i|$, so that $||\mu||_{\mathrm{Var}} = 1$ for any probability measure μ;

(b) for a signed measure μ and a probability transition matrix P, define the sign measure $(\mu P)_i = \sum_k \mu_k p_{ki}$. Then $||\mu P||_{\mathrm{Var}} \leqslant ||\mu||_{\mathrm{Var}}$;

(c) for a probability measure μ and a continuous-time Markov chain $P(t)$, $||\mu P(t)||_{\mathrm{Var}}$ is decreasing in $t \geqslant 0$.

(33) Consider the birth-death process with $b_n = \alpha n + \beta, a_n = \delta n, \alpha, \beta, \delta > 0$. Prove

(a) the process is recurrent iff $\alpha < \delta$ or $\alpha = \delta > \beta$;

(b) the process is ergodic iff $\alpha < \delta$;

(c) in other cases, the process is transient.

Hint: The proof will use the following Kummer's lemma:
Assume sequences $u_n > 0, v_n > 0$ such that $\sum 1/v_n = \infty$. Let the limit $\kappa = \lim_{n \to \infty} v_n u_n / n_{n+1} - v_{n+1}$. Then series $\sum_n u_n$ converges or diverges according to $\kappa > 0$ or $\kappa < 0$.

(34) Study the uniqueness, recurrence and ergodicity for the following single birth processes.

(a) Let $q_{i,i+1} = 1, q_{i,i-2} = 1 (i \geqslant 2), q_{10} = 1$, for other $i \neq j, q_{ij} = 0$.

(b) Let

$$q_{ij} = \begin{cases} a + \lambda i, & j = i + 1; \\ diq^{i-1}, & j = 0; \\ dipq^{i-j-1}, & j = 1, 2, \cdots, i - 1; \\ -(a + (\lambda + d)i), & j = i; \\ 0, & \text{else } i \neq j, \end{cases}$$

where $a > 0, \lambda > 0, p + q = 1$.

(35) Use the fact $r_i = \lim_{t \to \infty} p_{i0}(t) = \lim_{n \to \infty} \bar{p}_{i0}^{(n)}$ to prove Theorem 2.28, where $(\bar{p}_{i0}^{(n)})$ is the embedded chain.

(36) (Quasi-birth-death process) Consider two-dimensional Markov process $\{X_t, J_t\}$ on state space

$$E = \{(k,j) : k \geqslant 0, j = 1, 2, \cdots, m\}, \quad m < \infty.$$

Its Q-matrix is block-like:

$$Q = \begin{pmatrix} A_0 & C_0 & & & \\ B_1 & A & C & & \\ & B & A & C & \\ & & B & A & C \\ & & & \cdots & \cdots & \cdots \end{pmatrix},$$

where the sub-matrices of Q-matrix are of order m. Let R be the minimal nonnegative solution for matrix equation $R^2 B + RA + C = 0$.

(a) Prove the process is ergodic iff the spectral radium $\mathrm{sp}(R)$ is less than 1.

(b) Linear equation

$$\Pi_0(A_0 + RB_1) = 0, \quad \Pi_0(I - R)^{-1}e = 1$$

has a unique positive solution.

(c) Prove that the stationary distribution $\Pi = (\Pi_0, \Pi_1, \cdots)$ can be expressed as $\Pi_k = \Pi_0 R^k, k \geqslant 0$, where $\Pi_k = (\pi_{k1}, \cdots, \pi_{km})$.

Hint: Use Theorem 2.14 to derive $\Pi_{k-1}C + \Pi_k A + \Pi_{k+1}B = 0, k \geqslant 1$.

(37) (Reaction-diffusion process) Consider a Q-process on state space $\mathbb{Z}_+^S$ with S is finite or denumerable. For $u \in S, x \in E$, $x(u)$ denotes the number of particles at site u.

$$q(x,y) = \begin{cases} \lambda_0 + \lambda_1 x(u), & y = x + e_u; \\ \lambda_2 x(u), & y = x - e_u; \\ x(u)\rho(u,v), & y = x - e_u + e_v, u \neq v; \\ 0, & \text{else } y \neq x, \end{cases}$$

where $e_1 = (1, 0, \cdots)$, $e_2 = (0, 1, \cdots)$, $\cdots$ and $\rho(u, v)$ is a transition probability matrix on S with $\rho(u, u) = 0$. Study its uniqueness, recurrence and ergodicity.

Chapter 3

Reversible Markov Chains

3.1 Reversible and Symmetrizable Markov Chains

Definition 3.1. A Markov chain $(X_t)_{t \geqslant 0}$ is called **reversible**, if for every $n \geqslant 2, 0 \leqslant t_1 < t_2 < \cdots < t_n$ and $i_1, \cdots, i_n$, whenever

$$t_n - t_{n-1} = t_2 - t_1, \quad t_{n-1} - t_{n-2} = t_3 - t_2, \quad \cdots$$

it holds that

$$\mathbb{P}[X_{t_1} = i_1, \cdots, X_{t_n} = i_n] = \mathbb{P}[X_{t_1} = i_n, \cdots, X_{t_n} = i_1].$$

That is, the time reversal makes no difference on the distribution of the process. The film can be televised in the same way as it is televised in the reverse direction. This has significative physical meaning, and it is called **detailed balance** in statistical physics.

Specially, if the initial distribution is (π_i), then by

$$\mathbb{P}[X_0 = i, X_t = j] = \mathbb{P}[X_0 = j, X_t = i]$$

we have

$$\pi_i p_{ij}(t) = \pi_j p_{ji}(t). \tag{3.1}$$

Definition 3.2. $P(t)$ is called **reversible** with respect to π, if (3.1) holds for any t. Q is called **reversible** with respect to π, if

$$\pi_i q_{ij} = \pi_j q_{ji} \tag{3.2}$$

holds.

Theorem 3.3 (Existence theorem 1970's). $P^{\min}(t)$ is reversible with respect to π iff so is Q.

Theorem 3.4. The reversible Q-process is unique iff Q is reversible and the Q-process is unique (maybe dishonest).

Making summation in i on both sides of (3.1), we have

$$\sum_i \pi_i p_{ij}(t) = \pi_j \sum_i p_{ji}(t) = \pi_j,$$

provided $\sum_j p_{ij}(t) = 1$ (honest). Then the reversible measure is surely the stationary distribution. A mathematical generalization of the reversible one is to replace (π_i) by a sequence of positive numbers, which may not be summable. In this case, we will called $P(t)$ **symmetrizable**, provided (3.1) holds. This *distinguishes* the reversibility. For the symmetrizable process, it may not have the stationary distribution, or it is even not recurrent. Therefore we have two problems to study.

(1) Given Q, when is it symmetrizable and how can one find its symmetrizable measure?

(2) If a Q-process is symmetrizable, then when will it be recurrent?

To study the first problem, we introduced the tool of field theory; to study the second problem, researchers from UK introduced the tools of electric networks in 1980's. For an elegant introduction of the electric networks, refer to [21]. In book [10; Chapter 7], these were treated in a unified way, and one may find detailed answers to the two problems.

Consider the irreducible Markov chain $P(t)$. Let

$$w_{ij}(t) = \log p_{ij}(t) - \log p_{ji}(t), \qquad i, j \in E, \qquad t > 0.$$

We call $V(t) = (v_i(t))$ a **potential function** of $P(t)$, if

$$w_{ij}(t) = v_i(t) - v_j(t), i, j \in E, t > 0.$$

And $V(t)$ is also called a **potential field** of $P(t)$.

Theorem 3.5 (Basic theorem in field theory). The following statements are equivalent.

(1) $P(t)$ is symmetrizable.

(2) $P(t)$ is a potential field.

(3) $P(t)$ is conservative, in the sense that there exists a potential function of $P(t)$, which is independent of t.

Part (3) of the previous theorem and Theorem 2.13 say that the study on the symmetrization of $P^{\min}(t)$ can be reduced to that of its Q-matrix. Now, rewrite

$$w_{ij}(t) = \log \frac{p_{ij}(t)}{t} - \log \frac{p_{ji}(t)}{t}.$$

By letting $t \downarrow 0$, it follows that the corresponding time-independent difference of the potential from i to $j (\neq i)$ should be

$$w_{ij} = \log q_{ij} - \log q_{ji},$$

provided $q_{ij} > 0$ and $q_{ji} > 0$. The use of the potential fields reduces the problem to check the minimal closed path in the fields, and furthermore produces rather simple criterion for the symmetrizability. Refer to [10; §7.2].

To study the recurrence of $P(t)$, by Theorem 2.13, we only need consider the recurrence for the embedded chains. Using the method of electric networks, we have the following criterion.

Theorem 3.6 (Criterion for recurrent Markov chain). Suppose the irreducible Q-process has a symmetrizable measure π and let $a_{ij} = \pi_i q_{ij}/q_i$ for $i \neq j$. If $a_{ij} > 0$, then we draw a line between i and j, assigning it with electric resistance $1/a_{ij}$. Thus we get an electric network. Then Q is recurrent iff the effective resistance is infinite between every i and the infinity (by letting j go outside any finite network).

For the proof of the theorem, refer to [10; Theorem 7.19].

3.2   Estimate of Spectral Gap

Consider a matrix $Q = (q_{ij})$ on a denumerable set E such that $q_{ij} \geqslant 0$ $(i \neq j)$, $0 < q_i := -q_{ii} = \sum_{j \neq i} q_{ij} < \infty$. Recall, by reversibility, there is a probability measure $(\pi_i > 0 : i \in E)$ such that $\pi_i q_{ij} = \pi_j q_{ji}$ for all i, j. Then the corresponding operator $\Omega f(i) := \sum_j q_{ij}(f_j - f_i)$ $(i \in E)$ (pointwise on the class of functions having finite supports) is symmetric on $L^2(\pi)$ (the space of real square integrable functions). Let $P(t) = (p_{ij}(t))$ be the Q-process uniquely determined by Q. Then it is not difficult to prove that $P(t)$ is **strongly continuous** in $L^2(\pi)$:
$$\|P(t)f - f\| \to 0 \qquad \text{as } t \to 0,$$
where $\| \cdot \|$ denotes the $L^2(\pi)$-norm. And $P(t)$ is **strongly contractive**: $\|P(t)f\| \leqslant \|f\|$ holds for every t. Let Ω be the strong infinitesimal generator of $P(t)$ in $L^2(\pi)$ (coincides with the (point-wise) operator Ω above on the functions with finite supports but they may have different domains), and denote its domain by $\mathscr{D}(\Omega)$.

What we concerns is the following L^2-**exponential convergence**:
$$\|P(t)f - \pi(f)\| \leqslant \|f - \pi(f)\|e^{-\varepsilon t}, \qquad t \geqslant 0, \ f \in L^2(\pi),$$
where $\varepsilon > 0$ is positive constant.

In the following, we are going to describe the maximal convergence rate $\varepsilon_{\max}$. For this, define the **spectral gap** of operator Ω
$$\mathrm{gap}(\Omega) = \inf\{-(\Omega f, f) : \pi(f) = 0, \|f\| = 1\},$$

where $(f, g) = \sum_i \pi_i f_i g_i$ and $\pi(f) = \sum_i \pi_i f_i$ for $f, g \in L^2(\pi)$. To see the nonnegativeness of gap(Ω), we need only

$$-(f, \Omega f) = -\sum_i \pi_i f_i \sum_j q_{ij} f_j = \frac{1}{2} \sum_{i,j} \pi_i q_{ij} (f_j - f_i)^2 \geqslant 0.$$

Next, by the basic spectral theory (cf. Appendix in this chapter),

$$\left(\frac{f - P(t)f}{t}, f \right) \uparrow \text{ some } D(f, f) \leqslant +\infty, \quad \text{as } t \downarrow 0. \tag{3.3}$$

And

$$D(f, f) = \frac{1}{2} \sum_{i,j} \pi_i q_{ij} (f_j - f_i)^2.$$

The uniqueness of the process ensures that the functions, which make $D(f, f)$ finite, constitutes the domain:

$$\mathscr{D}(D) = \{f \in L^2(\pi) : D(f, f) < \infty\} \qquad \text{(This proof is difficult).}$$

Now define

$$\text{gap}(D) = \inf\{D(f, f) : \pi(f) = 0 \text{ and } \pi(f^2) = 1\}.$$

Having these notations, we can state our basic theorem on convergence rate.

Theorem 3.7. $\varepsilon_{\max} = \text{gap}(\Omega) = \text{gap}(D)$.

Proof Since $D(f, f) = -(\Omega f, f)$ for $f \in \mathscr{D}(\Omega) \subset \mathscr{D}(D)$, we have $\text{gap}(D) \leqslant \text{gap}(\Omega)$. To prove $\varepsilon_{\max} \geqslant \text{gap}(\Omega)$, we need only use the following fact

$$\frac{\mathrm{d}}{\mathrm{d}t} \|P(t)f\|^2 = 2(P(t)f, \Omega P(t)f) \leqslant -2 \, \text{gap}(\Omega) \|P(t)f\|^2,$$

$$f \in \mathscr{D}(\Omega), \ \pi(f) = 0, \ \|f\| = 1,$$

and the density of $\mathscr{D}(\Omega)$ in $L^2(\pi)$. Finally, let $f \in \mathscr{D}(D)$ satisfy $\pi(f) = 0$ and $\|f\| = 1$. Then

$$D(f, f) = \lim_{t \downarrow 0} \frac{1}{t}(f - P(t)f, f) \geqslant \lim_{t \downarrow 0} \frac{1}{t}(1 - e^{-\varepsilon t}) = \varepsilon.$$

Thus $\text{gap}(D) \geqslant \varepsilon$. $\square$

As in the finite-dimensional case (see Appendix in this chapter), we write gap(D) by λ_1.

Now, we define the **graph structure** associated to matrix $Q = (q_{ij})$. A pair $\langle i, j \rangle$ $(i \neq j)$ is called an edge, if $q_{ij} > 0$. The connecting edges $\langle i, i_1 \rangle, \langle i_1, i_2 \rangle, \cdots, \langle i_n, j \rangle$ (i, j and each i_k are different) constitute a path from i to j. Assume for every pair $i \neq j$, there exists a path from i to j.

(This is equivalent to the irreducibility of Q.) Choose and fix such a path γ_{ij}. Next, assign on each edge $e = \langle i, j \rangle$ a positive weight $\{w(e)\}$, and let $|\gamma_{ij}|_w = \sum_{e \in \gamma_{ij}} w(e)$. For $e = \langle i, j \rangle$, set $a(e) = \pi_i q_{ij}$ and

$$I(w)(e) = \frac{1}{a(e)w(e)} \sum_{\{i,j\}: \gamma_{ij} \ni e} |\gamma_{ij}|_w \pi_i \pi_j,$$

where $\{i, j\}$ denotes the unordered pair of i and j. Here is a simple lower bound of λ_1.

Theorem 3.8. We have $\lambda_1 \geqslant \sup_{w \in \mathscr{W}} \inf_e I(w)(e)^{-1}$, where $\mathscr{W}$ is the set of weights $\{w(e)\}$.

Proof For the sake of simplicity, let $f(e) = f_j - f_i$ for $e = \langle i, j \rangle$. The Cauchy-Schwarz inequality implies

$$(f_i - f_j)^2 = \left(\sum_{e \in \gamma_{ij}} f(e) \right)^2 \leqslant \left(\sum_{e \in \gamma_{ij}} \frac{f(e)^2}{w(e)} \right) |\gamma_{ij}|_w.$$

Thus, for each f with $\pi(f) = 0$ and $\pi(f^2) = 1$, we have

$$1 = \frac{1}{2} \sum_{i,j} \pi_i \pi_j (f_i - f_j)^2 = \sum_{\{i,j\}} \pi_i \pi_j \left(\sum_{e \in \gamma_{ij}} f(e) \right)^2$$

$$\leqslant \sum_{\{i,j\}} \pi_i \pi_j \left(\sum_{e \in \gamma_{ij}} \frac{f(e)^2}{w(e)} \right) |\gamma_{ij}|_w$$

$$= \sum_e a(e) f(e)^2 \frac{1}{a(e)w(e)} \sum_{\{i,j\}: \gamma_{ij} \ni e} |\gamma_{ij}|_w \pi_i \pi_j$$

$$\leqslant D(f, f) \sup_e I(w)(e). \qquad \square$$

From the proof, we see that the use of the graph structure is rather natural, since only the pair $\{i, j\}$ with $q_{ij} > 0$ can appear in Dirichlet form $D(f, f)$. However there are examples that show this graph structure is not always necessary. Of course, in practice the key is how to choose $\{w(e)\}$, this is more important for infinite E. The following variational formula is another example.

Theorem 3.9. Consider a birth-death process on $E = \{0, 1, 2, \cdots, N\}$ with $N \leqslant \infty$. Its Q-matrix is $q_{i,i+1} = b_i > 0$ $(0 \leqslant i < N)$, $q_{i,i-1} = a_i > 0$ $(1 \leqslant i < N + 1)$ and $q_{ij} = 0$ for other $i \neq j$. Let $\mathscr{W}$ be the set of all strictly increasing sequence (w_i) such that $\sum_{i=0}^{N} \mu_i w_i \geqslant 0$ and define

$$I_i(w) = \frac{1}{b_i \mu_i (w_{i+1} - w_i)} \sum_{j=i+1}^{N} \mu_j w_j, \qquad 0 \leqslant i < N,$$

where

$$\mu_0 = 1, \quad \mu_n = \frac{b_0 \cdots b_{n-1}}{a_1 \cdots a_n}, \quad 1 \leqslant n < N+1.$$

Then

$$\lambda_1 = \sup_{w \in \mathscr{W}} \inf_{0 \leqslant i \leqslant N-1} I_i(w)^{-1}. \tag{3.4}$$

Proof a) Note that we can replace all $\mu's$ in $I_i(w)$ by the stationary distribution π, where $\pi_i = \mu_i/\mu$ $(0 \leqslant i < N+1)$ with $\mu := \sum_{j=0}^{N} \mu_j$. Let e_i denote the edge $\langle i, i+1 \rangle$. Obviously, for each pair $i < j$, there exists a unique path (no loop), which is constituted by $e_i, e_{i+1}, \cdots, e_{j-1}$. Take $w(e_i) = w_{i+1} - w_i$. Then

$$|\gamma_{k\ell}|_w = (w_{k+1} - w_k) + \cdots + (w_\ell - w_{\ell-1}) = w_\ell - w_k.$$

Thus

$$\sum_{\{k,\ell\}:\gamma_{k\ell} \ni e_i} |\gamma_{k\ell}|_w \pi_k \pi_\ell = \sum_{k=0}^{i} \sum_{\ell=i+1}^{N} \pi_k \pi_\ell (w_\ell - w_k)$$

$$= \sum_{k=0}^{i} \pi_k \sum_{\ell=i+1}^{N} \pi_\ell w_\ell - \sum_{k=0}^{i} \pi_k w_k \sum_{\ell=i+1}^{N} \pi_\ell$$

$$= \sum_{\ell=i+1}^{N} \pi_\ell w_\ell - \left(\sum_{k=i+1}^{N} \pi_k \right) \sum_{\ell=i+1}^{N} \pi_\ell w_\ell - \sum_{k=0}^{i} \pi_k w_k \sum_{\ell=i+1}^{N} \pi_\ell$$

$$= \sum_{\ell=i+1}^{N} \pi_\ell w_\ell - \left(\sum_{k=i+1}^{N} \pi_k \right) \left(\sum_{\ell=0}^{N} \pi_\ell w_\ell \right)$$

$$\leqslant \sum_{\ell=i+1}^{N} \pi_\ell w_\ell, \quad 0 \leqslant i < N.$$

By Theorem 3.8, we have proved the inequality:

$$\lambda_1 \geqslant \sup_{w \in \mathscr{W}} \inf_{0 \leqslant i \leqslant N-1} I_i(w)^{-1}.$$

b) The proof of equality is much more difficult, so we have to use some properties of the eigenfunction.

Lemma 3.10. Suppose $\lambda > 0$ and $g \not\equiv 0$ is a solution of equation $\Omega g = -\lambda g$. Then $g_0 \neq 0$ and

$$\pi_n b_n (g_{n+1} - g_n) = -\lambda \sum_{i=0}^{n} \pi_i g_i, \quad 0 \leqslant n < N+1. \tag{3.5}$$

Here by convention $b_N = 0$ if $N < \infty$.

Proof By convention, let $a_0 = 0$, and $a_{N+1} = 0$ if $N < \infty$. Since $\pi_i b_i = \pi_{i+1} a_{i+1}$ for $0 \leqslant i < N + 1$, we have

$$-\lambda \sum_{i=0}^{n} \pi_i g_i = \sum_{i=0}^{n} \pi_i \Omega g(i) = \sum_{i=0}^{n} \left[\pi_i a_i (g_{i-1} - g_i) + \pi_i b_i (g_{i+1} - g_i) \right]$$

$$= \sum_{i=0}^{n} \left[-\pi_i a_i (g_i - g_{i-1}) + \pi_{i+1} a_{i+1} (g_{i+1} - g_i) \right]$$

$$= -\pi_0 a_0 (g_0 - g_{-1}) + \pi_{n+1} a_{n+1} (g_{n+1} - g_n)$$

$$= \pi_n b_n (g_{n+1} - g_n)$$

where the term g_{-1} can be omitted since $a_0 = 0$.

If $g_0 = 0$, then by (3.5), we have inductively that $g_i \equiv 0$. This contradicts the assumption $g \not\equiv 0$. □

Lemma 3.11. Assume $\lambda_1 > 0$ and g is a solution of equation $\Omega g = -\lambda_1 g$ satisfying $g_0 < 0$. Then g is strictly increasing.

Proof For convenience, set $a_0 = 0$, and $b_N = 0$ (if $N < \infty$). Since $g_0 < 0$ and $\pi_0 b_0 (g_1 - g_0) = -\pi_0 g_0$, we have $g_1 > g_0$. Now suppose there exists some n, $1 \leqslant n < N$ such that

$$g_0 < g_1 < \cdots < g_{n-1} < g_n \geqslant g_{n+1}. \tag{3.6}$$

We will prove this can not happen.

By (3.5), we have for $0 \leqslant k < N$,

$$g_k < (\text{respectively, } =) \, g_{k+1} \iff \sum_{i=0}^{k} \pi_i g_i < (\text{respectively, } =) \, 0.$$

Set $\tilde{g}_n = -\sum_{i=0}^{n-1} \pi_i g_i / \pi_n$. By (3.5) again, we have

$$g_k < (\text{respectively, } =) \, g_{k+1} \iff \sum_{i=0}^{k} \pi_i g_i < (\text{respectively, } =) \, 0.$$

This, together with (3.5) and (3.6), implies

$$\sum_{i \leqslant n-1} \pi_i g_i + \pi_n \tilde{g}_n = 0 \tag{3.7}$$

and

$$g_n \geqslant \tilde{g}_n = \frac{\pi_{n-1} b_{n-1}}{\lambda_1 \pi_n} (g_n - g_{n-1}) = \frac{a_n}{\lambda_1} (g_n - g_{n-1}) > 0. \tag{3.8}$$

Set $\bar{g}_i = g_i I_{[i < n]} + g_n I_{[i \geqslant n]}$. Then

$$\sum_i \pi_i \bar{g}_i^2 = \sum_{i \leqslant n-1} \pi_i g_i^2 + g_n^2 \sum_{i=n}^{N} \pi_i$$

and

$$\sum_i \pi_i \bar{g}_i = \sum_{i \leqslant n-1} \pi_i g_i + g_n \sum_{i=n}^N \pi_i = g_n \sum_{i=n}^N \pi_i - \pi_n \tilde{g}_n. \qquad \text{(by (3.7))}$$

Thus

$$\sum_i \pi_i \bar{g}_i^2 - \left(\sum_i \pi_i \bar{g}_i\right)^2 = \sum_{i \leqslant n-1} \pi_i g_i^2 + g_n^2 \sum_{i=n}^N \pi_i - \left(g_n \sum_{i=n}^N \pi_i - \pi_n \tilde{g}_n\right)^2. \quad (3.9)$$

Note that

$$-\sum_i \pi_i (\bar{g}\Omega\bar{g})(i) = \lambda_1 \sum_{i \leqslant n-1} \pi_i g_i^2 + \pi_n a_n g_n (g_n - g_{n-1})$$

$$= \lambda_1 \sum_{i \leqslant n-1} \pi_i g_i^2 + \lambda_1 \pi_n g_n \tilde{g}_n. \qquad \text{(by (3.8))}$$

$$(3.10)$$

We now prove

$$\pi_n g_n \tilde{g}_n < g_n^2 \sum_{i=n}^N \pi_i - \left(g_n \sum_{i=n}^N \pi_i - \pi_n \tilde{g}_n\right)^2. \qquad (3.11)$$

By (3.8), we have $g_n > 0$. Thus (3.11) is equivalent to

$$\pi_n \frac{\tilde{g}_n}{g_n} < \sum_{i=n}^N \pi_i - \left(\sum_{i=n}^N \pi_i - \pi_n \frac{\tilde{g}_n}{g_n}\right)^2,$$

or

$$\left(\sum_{i=n}^N \pi_i - \pi_n \frac{\tilde{g}_n}{g_n}\right)^2 < \sum_{i=n}^N \pi_i - \pi_n \frac{\tilde{g}_n}{g_n}.$$

The last inequality comes from that $0 < \tilde{g}_n \leqslant g_n$, and

$$0 < \sum_{i=n}^N \pi_i - \pi_n \tilde{g}_n/g_n \leqslant \sum_{i=n}^N \pi_i < 1.$$

This proves (3.11).

Summing up (3.9)–(3.11), we obtain

$$\lambda_1 \leqslant \frac{-\sum_i \pi_i (\bar{g}\,\Omega\,\bar{g})(i)}{\sum_i \pi_i \bar{g}_i^2 - \left(\sum_i \pi_i \bar{g}_i\right)^2}$$

$$= \frac{\lambda_1 \sum_{i \leqslant n-1} \pi_i g_i^2 + \lambda_1 \pi_n g_n \tilde{g}_n}{\sum_{i \leqslant n-1} \pi_i g_i^2 + g_n^2 \sum_{i=n}^N \pi_i - \left(g_n \sum_{i=n}^N \pi_i - \pi_n \tilde{g}_n\right)^2} < \lambda_1.$$

This is a contradiction.   □

Now, let us come back to prove Theorem 3.9. To obtain the equality in (3.4), note that if $\lambda_1 = 0$, then the previous proof of a) shows the equality holds. Therefore, in the following we assume that $\lambda_1 > 0$.

c) By Lemma 3.11, we can define a sequence of positive numbers $u_i = g_{i+1} - g_i$, $0 \leqslant i < N$. For convenience, define further $u_N = 1$. Then by eigen-equation $\Omega g = -\lambda_1 g$, we have

$$b_i u_i - a_i u_{i-1} = -\lambda_1 g_i \quad (a_0 := 0), \qquad 0 \leqslant i < N. \tag{3.12}$$

Replacing i with $i + 1$ in (3.12) gives another equation. Making difference of these two equations, we have

$$R_i(u) := (a_{i+1} u_i - b_{i+1} u_{i+1} - a_i u_{i-1} + b_i u_i)/u_i = \lambda_1 > 0, \qquad 0 \leqslant i < N.$$

By (3.5) and Lemma 3.11, there exists nonnegative limit $c := \lim_{n \to \infty} \mu_n b_n u_n$ provided $N = \infty$. Otherwise $c := \mu_N b_N u_N = 0$. By convention, $u_{-1} = 0$. From (3.12) and the monotonicity of g_i, we have

$$\mu_n b_n u_n = -\lambda_1 \sum_{i=0}^{n} \mu_i g_i \leqslant -\lambda_1 \sum_{i \leqslant n : g_i < 0} \mu_i g_i$$

$$\leqslant -\lambda_1 g_0 \sum_{i \leqslant n : g_i < 0} \mu_i \leqslant -\lambda_1 g_0 Z < \infty.$$

Thus $c < \infty$, further $g \in L^1(\pi)$.

Let

$$w_i = a_i u_{i-1} - b_i u_i + c/(\mu - \mu_0) = \lambda_1 g_i + c/(\mu - \mu_0).$$

Then

$$(w_{i+1} - w_i)/u_i = R_i(u) = \lambda_1 > 0, \qquad 0 \leqslant i < N. \tag{3.13}$$

So w_i is strictly increasing. On the other hand, notice

$$\sum_{j=i+1}^{N} \mu_j w_j = \sum_{j=i+1}^{N} [\mu_j a_j u_{j-1} - \mu_j b_j u_j + c\mu_j/(\mu - \mu_0)]$$

$$= \sum_{j=i+1}^{N} [\mu_{j-1} b_{j-1} u_{j-1} - \mu_j b_j u_j + c\mu_j/(\mu - \mu_0)]$$

$$= \mu_i b_i u_i - c + \frac{c}{\mu - \mu_0} \sum_{j=i+1}^{N} \mu_j$$

$$= b_i \mu_i u_i - \frac{c}{\mu - \mu_0} \sum_{1 \leqslant j \leqslant i} \mu_j \leqslant b_i \mu_i u_i, \qquad 0 \leqslant i < N. \tag{3.14}$$

We have in particular that $\sum_{j\geqslant 1}^{N} \mu_j w_j = \mu_0 b_0 u_0 > 0$ and $w \in L^1(\pi)$.

d) Now we claim that $\sum_{j=i+1}^{N} \mu_j w_j > 0$ for every $0 \leqslant i < N$. Indeed, if $w_i \geqslant 0$ for $0 \leqslant i < N$, then by the increasing monotonicity of w, we get the conclusion. Otherwise, if there exists i such that $w_i < 0$, then

$$w_{i-1}, \cdots, w_1 < 0,$$

so that

$$\sum_{j=i+1}^{N} \mu_j w_j = \sum_{j=1}^{N} \mu_j w_j - \sum_{j=1}^{i} \mu_j w_j > 0.$$

Next by $w_0 = -b_0 u_0 + c/(\mu - \mu_0)$, we have

$$\sum_{j} \mu_j w_j = w_0 + \sum_{j=1}^{N} \mu_j w_j = c/(\mu - \mu_0) \geqslant 0.$$

Thus $w \in \mathscr{W}$. Summing up these facts derives

$$I_i(w)^{-1} = b_i \mu_i R_i(u) u_i \Big/ \sum_{j\geqslant i+1} \mu_j w_j \geqslant R_i(u) = \lambda_1, \quad 0 \leqslant i < N. \quad (3.15)$$

Furthermore, $\inf_{0\leqslant i<N} I_i(w)^{-1} \geqslant \lambda_1$. From this and part a) above, we have

$$\inf_{0\leqslant <N} I_i(w)^{-1} = \lambda_1.$$

Hence, we get the equality. Thus the equality in (3.15) holds, and so does the one in (3.14). This implies $c = 0$.   □

During the proof, we also obtain the following conclusion:

Lemma 3.12. For g given by Lemma 3.11, we have $\pi(g) = 0$.

To contrast spectral gap with exponential ergodicity, we have the following result, whose proof omitted here. Cf. [10; Theorem 9.15]. Recall a jump process $P(t) = (p_{ij}(t))$ with stationary distribution π is **exponentially ergodic** iff for any i, j, there exist $C_{ij} < \infty$ and $\lambda > 0$ such that

$$|p_{ij}(t) - \pi_j| \leqslant C_{ij} e^{-\lambda t}.$$

The optimal λ is called the **exponential convergence rate**.

Theorem 3.13. For the reversible Markov chain, the exponential convergence rate equals the spectral gap.

Now, we turn to consider the Dirichlet spectral gaps for jump processes. Fix a point, say $0 \in E$. Then Dirichlet eigenvalue is defined as

$$\lambda_0 = \inf\{D(f, f) : f(0) = 0 \text{ and } \pi(f^2) = 1\}.$$

For each $i \in E$, we assume and choose a path from 0 to i (with no loop) γ_i. And on each edge, assign a positive weight function $\{w(e)\}$. Let $|\gamma_i|_w = \sum_{e \in \gamma_i} w(e)$ and

$$\tilde{I}(w)(e) = \frac{1}{a(e)w(e)} \sum_{i:\, \gamma_i \ni e} |\gamma_i|_w \pi_i.$$

Theorem 3.14. We have $\lambda_0 \geqslant \sup\limits_{w} \inf\limits_{e} \tilde{I}(w)(e)^{-1}$.

Proof

$$1 = \sum_i \pi_i f_i^2 = \sum_i \pi_i (f_i - f_0)^2 = \sum_i \pi_i \left(\sum_{e \in \gamma_i} f(e) \right)^2$$

$$\leqslant \sum_i \pi_i \sum_{e \in \gamma_i} \frac{f(e)^2}{w(e)} |\gamma_i|_w = \sum_e a(e) f(e)^2 \tilde{I}(w)(e)$$

$$\leqslant D(f, f) \sup_e \tilde{I}(w)(e). \qquad \square$$

Theorem 3.15. Assume that (a_i, b_i), (μ_i), $(I_i(w))$ as in Theorem 3.9 but let $\mathscr{W}_0$ be all strictly increasing sequences (w_i) with $w_0 = 0$. Then $\lambda_0 = \sup\limits_{w \in \mathscr{W}} \inf\limits_{0 \leqslant i < N} I_i(w)^{-1}$.

When $b_0 = 0$, if change the definition

$$I_i(w) = \frac{1}{a_{i+1}\tilde{\mu}_{i+1}(w_{i+1} - w_i)} \sum_{j=i+1}^{N} \tilde{\mu}_j w_j,$$

where

$$\tilde{\mu}_1 = 1, \quad \tilde{\mu}_n = \frac{b_1 \cdots b_{n-1}}{a_1 \cdots a_n}, \qquad n \geqslant 2,$$

then the assertion remains true.

Proof a) Let again e_i be the edge of $\langle i, i+1 \rangle$. For each $i \geqslant 1$, there exists a path which is constituted by $e_0, e_1, \cdots, e_{i-1}$. Let $w(e_i) = w_{i+1} - w_i$. Then

$$\sum_{k:\, \gamma_k \ni e_i} |\gamma_k|_w \pi_k = \sum_{k=i+1}^{N} (w_k - w_0)\pi_k = \sum_{k=i+1}^{N} \pi_k w_k.$$

Thus

$$\tilde{I}(w)(e_i) = \frac{1}{\pi_i b_i (w_{i+1} - w_i)} \sum_{j=i+1}^{N} \pi_i w_i = I_i(w),$$

where $I_i(w)$ is given in Theorem 3.9. From Theorem 3.14, we have proved the inequality.

b) The rest of proof of the theorem is similar to that of the second part of Theorem 3.9. For the sake of completeness, we still give the details. Let $\lambda_0 > 0$, and $g \not\equiv 0$, $g_0 = 0$ be a solution of equation $\Omega g(i) = -\lambda_0 g_i$ ($1 \leqslant i < N + 1$). Here, by convention, $a_0 = 0$ and $b_N = 0$. The key to prove the equality is to demonstrate that (g_i) is strictly increasing. Once this was proved, without loss of generality, we may assume $g_i \uparrow$. Since $\mu_i b_i = \mu_{i+1} a_{i+1}$, we have

$$I_i(g) = \frac{1}{\mu_{i+1} a_{i+1}(g_{i+1} - g_i)} \sum_{j=i+1}^{N} \mu_j g_j \equiv \frac{1}{\lambda_0} \qquad (3.16)$$

holds for all $0 \leqslant i < N$. Thus we obtain the desired assertion.

c) To prove (3.16), we first note that

$$-\lambda_0 \sum_{1}^{n} \pi_i g_i = \sum_{1}^{n} \pi_i \Omega g(i) = \sum_{1}^{n} \left[\pi_i a_i (g_{i-1} - g_i) + \pi_i b_i (g_{i+1} - g_i) \right]$$

$$= \sum_{1}^{n} \left[-\pi_i a_i (g_i - g_{i-1}) + \pi_{i+1} a_{i+1}(g_{i+1} - g_i) \right] \qquad (3.17)$$

$$= \pi_{n+1} a_{n+1}(g_{n+1} - g_n) - \pi_1 a_1 g_1.$$

Using the assumption $g_i \uparrow$ and mimicking the proof of b) in Theorem 3.9, we can derive

$$\lim_{n \to \infty} \pi_{n+1} a_{n+1}(g_{n+1} - g_n) = 0$$

if $N = \infty$, and $\pi_N a_{N+1}(g_{N+1} - g_N) = 0$ if $N < \infty$. In particular, $\lambda_0 \sum_{1}^{N} \pi_i g_i = \pi_1 a_1 g_1$. Thus

$$\lambda_0 \sum_{i=n+1}^{N} \pi_i g_i = \pi_{n+1} a_{n+1}(g_{n+1} - g_n),$$

which implies (3.16).

d) Now we turn back to prove the strict monotonicity of the eigenfunction (g_i) for λ_0. By (3.17), we have $g_1 \neq 0$. Otherwise by induction it implies that $g_i \equiv 0$, $i \geqslant 1$. Thus, we may assume $g_1 > 0$. Suppose there is some n: $1 \leqslant n < N$ such that

$$0 = g_0 < g_1 < \cdots < g_{n-1} < g_n \geqslant g_{n+1}.$$

Define $\bar{g}_i = g_i I_{[i<n]} + g_n I_{[i \geqslant n]}$. Then

$$\sum_i \pi_i \bar{g}_i^2 = \sum_{i \leqslant n-1} \pi_i g_i^2 + g_n^2 \sum_{i=n}^{N} \pi_i,$$

$$-\sum_i \pi_i (\bar{g} \Omega \bar{g})(i) = \lambda_0 \sum_{i \leqslant n-1} \pi_i g_i^2 + \pi_n a_n g_n (g_n - g_{n-1}).$$

Since

$$\lambda_0 g_n = -\Omega g(n) = b_n(g_n - g_{n+1}) + a_n(g_n - g_{n-1}) \geqslant a_n(g_n - g_{n-1}),$$

we have

$$\pi_n a_n g_n (g_n - g_{n-1}) \leqslant \lambda_0 \pi_n g_n^2 < \lambda_0 g_n^2 \sum_{i=n}^{N} \pi_i.$$

Therefore, by the definition of λ_0,

$$\lambda_0 \leqslant \frac{-\sum_i \pi_i (\bar{g}\Omega\bar{g})(i)}{\sum_i \pi_i \bar{g}_i^2} = \frac{\lambda_0 \sum_{i \leqslant n-1} \pi_i g_i^2 + \pi_n a_n g_n (g_n - g_{n-1})}{\sum_{i \leqslant n-1} \pi_i g_i^2 + g_n^2 \sum_{i=n}^{N} \pi_i} < \lambda_0,$$

which is a contradiction.

e) As for the final assertion, noticing that in the proofs a)–d) above, we do not use π_0 and b_0 since $g_0 = 0$. Furthermore, the previous $I_i(w)$ is homogeneous in (μ_i). $\square$

3.3   Appendix: Spectral Representation of Reversible Markov Chains

Let $(p_{ij}(t))$ be a reversible Markov chain, namely, there exists a probability measure $(\pi_i : i \in E)$ such that

$$\pi_i p_{ij}(t) = \pi_j p_{ji}(t) \quad i,j \in E, \quad t \geqslant 0. \tag{3.18}$$

When the Q-process is unique, this is equivalent to

$$\pi_i q_{ij} = \pi_j q_{ji} \quad i,j \in E. \tag{3.19}$$

Since we have the measure (π_i) in advance, it is natural to consider the function space $L^2(\pi)$ constituted by real square integrable functions. It is not hard to prove that (3.18) is equivalent to

$$(P(t)f, g) = (f, P(t)g), \quad f, g \in L^2(\pi), \quad t \geqslant 0, \tag{3.20}$$

where $(\cdot, \cdot)$ denotes the usual inner product in $L^2(\pi)$ and $P(t) = (p_{ij}(t))$. In other words, $P(t)$ is a **self-adjoint operator** on $L^2(\pi)$ (similar to the symmetric matrix in a finite-dimensional space). For Markov chains, it is easy to check $P(t)$ is strongly continuous in $L^2(\pi)$:

$$\|P(t)f - f\| \to 0 \qquad \text{as } t \to 0,$$

where $\|\cdot\|$ denotes the L^2-norm in $L^2(\pi)$. $P(t)$ is further strongly contractive: $\|P(t)f\| \leqslant \|f\|$ holds for every t. Let Ω denote the strong infinitesimal operator of $P(t)$ in $L^2(\pi)$. Namely, there exists a dense subspace $\mathscr{D}(\Omega) \subset L^2(\pi)$ such that for any $f \in \mathscr{D}(\Omega)$, limit

$$\Omega f = \lim_{t \to 0} \frac{1}{t}(P_t f - f)$$

exists in $L^2(\pi)$. And $\mathscr{D}(\Omega)$ is called the domain of Ω.

Theorem 3.16 (Spectral representation theorem). $P(t)$ and Ω have the following representation:

$$P(t) = \int_0^\infty e^{-\lambda t}\mathrm{d}E_\lambda, \qquad \Omega = \int_0^\infty -\lambda \mathrm{d}E_\lambda, \tag{3.21}$$

where $\{E_\lambda : \lambda \geqslant 0\}$ is a family of increasing projection-valued measures on $L^2(\pi)$, that is, $E_0 = 0$ and $E_\infty = I$ (identity operator), for every $\lambda, \mu : 0 \leqslant \lambda \leqslant \mu \leqslant \infty, E_\lambda^2 = E_\lambda,\ E_\lambda E_\mu = E_\mu E_\lambda = E_\lambda$. And for each $f \in L^2(\pi)\phi(\lambda) := (E_\lambda f, f)$ is a finite measure on $[0, \infty)$ with respect to λ.

In more details, for instance the first term in (3.21) means

$$(P(t)f, g) = \int_0^\infty e^{-\lambda t}\mathrm{d}(E_\lambda f, g), \qquad f, g \in L^2(\pi),$$

where

$$(E_\lambda f, g) = \frac{1}{4}\big[(E_\lambda(f + g), f + g) - (E_\lambda(f - g), f - g)\big].$$

Thus, $(E_\lambda f, g)$ is a signal measure on $[0, \infty)$ for fixed f and g.

This theorem is fundamental in the spectral theory, but the beginner may not find it easy to understand. The purpose of the appendix is trying to explain how to deduce (3.21). Before discussing the problem, let us give a simple corollary of (3.21). Note that when t decreases,

$$(1 - e^{-\lambda t})/(\lambda t) = t^{-1}\int_0^t e^{-\lambda s}\mathrm{d}s$$

increases. Thus the limit

$$\left(\frac{f - P(t)f}{t}, f\right) \uparrow \text{ some } D(f, f) \leqslant +\infty \quad (\text{as } t \downarrow 0) \qquad (3.22)$$

always exists. The right-hand side of (3.22) is called the **Dirichlet form** for the process $P(t)$. Without the help of spectral theory (3.21), we do not know how to prove the increasing property of (3.22).

Of course, we begin with the simplest case. Assume that E is a finite set. In this case, the process $P(t)$ is unique, and has the following direct expression:

$$P(t) = e^{tQ} = \sum_{n=0}^{\infty} \frac{t^n}{n!} Q^n.$$

a) Without loss of generality, assume that $E = \{0, 1, 2, \cdots, N\}$. It follows from (3.18) that

$$\left(f - P(t)f, f\right) = \frac{1}{2} \sum_{i, j \in E} \pi_i p_{ij}(t)(f_j - f_i)^2.$$

From this and (3.22), we obtain

$$D(f, f) = \frac{1}{2} \sum_{i, j \in E} \pi_i q_{ij}(f_j - f_i)^2 = (-\Omega f, f) \qquad (3.23)$$

holds for every $f \in L^2(\pi)$.

b) From (3.23) we see that operator $-\Omega$ is positive definitive. In other words, $(\pi_i q_{ij})$ is a positive definitive matrix.

Recall that in the study of quadratic form or quadratic surface, we always transform it into the standard form. For this, we introduce the notations:

$$\Pi = \text{diag}(\pi_i) \quad (\text{diagonal matrix with elements } \pi_i),$$

$$A = \Pi^{1/2} Q \Pi^{-1/2},$$

$$f = (f_i) \quad (\text{row vector}).$$

Noting that $\Pi Q = Q^* \Pi$, from (3.23) we have

$$
\begin{aligned}
(-\Omega f, f) &= -f^* \Pi Q f = -f^* \Pi^{1/2} (\Pi^{1/2} Q \Pi^{-1/2}) \Pi^{1/2} f \quad (\text{since } \Omega = Q) \\
&= -(\Pi^{1/2} f)^* (\Pi^{1/2} Q \Pi^{-1/2})(\Pi^{1/2} f) \\
&= -(\Pi^{1/2} f)^* A (\Pi^{1/2} f).
\end{aligned}
$$

$$(3.24)$$

Since

$$
\begin{aligned}
A^* &= (\Pi^{1/2} Q \Pi^{-1/2})^* = \Pi^{-1/2} Q^* \Pi^{1/2} = \Pi^{-1/2} (\Pi Q) \Pi^{-1/2} \\
&= \Pi^{1/2} Q \Pi^{-1/2} = A,
\end{aligned}
$$

$$(3.25)$$

we see that A is symmetric. Thus there exists an orthogonal matrix U such that

$$-UAU^* = \text{diag}\,(\lambda_i),$$

where $0 \leqslant \lambda_0 \leqslant \lambda_1 \leqslant \cdots \leqslant \lambda_N$ are the eigenvalues of $-A$.

c) Next, we prove that both A and Q have the same eigenvalues. This can be seen from

$$Af = \rho f \Longleftrightarrow \Pi^{1/2}Q\Pi^{-1/2}f = \rho f$$
$$\Longleftrightarrow Q\Pi^{-1/2}f = \rho\Pi^{-1/2}f.$$

Now, from $Q1 = 0 = 0 \cdot 1$ we have $\lambda_0 = 0$. If Q is irreducible and E is finite, then $\lambda_1 > 0$. We call the difference $\lambda_1 - \lambda_0 = \lambda_1$ the **spectral gap** of operator Ω. This is just $\text{gap}(D)$ defined before.

d) From (3.24) and (3.25), it follows f that

$$(-\Omega f, f) = -(\Pi^{1/2}f)^*A(\Pi^{1/2}f) = (\Pi^{1/2}f)^*U^*\text{diag}\,(\lambda_i)U(\Pi^{1/2}f)$$
$$= \sum_{i=1}^{N} \lambda_i (U\Pi^{1/2}f)_i^2. \tag{3.26}$$

Now we take a key step forward. Define a finite measure $F_f(d\lambda)$ possessing weight $x_k := (U\Pi^{1/2}f)_k^2$ $(k = 0, 1, \cdots, N)$ at λ_k. Then (3.26) can be rewritten as

$$(-\Omega f, f) = \int_0^\infty \lambda F_f(d\lambda).$$

This integral form is necessary for infinite space, as the spectrum may be continuous and then the summation in (3.26) may be meaningless. In the infinite case, the orthogonal transform U is replaced by unitary transform and the domain of operator Ω is usually a dense subset of $L^2(\pi)$, not the space $L^2(\pi)$ itself.

Furthermore, we have

$$F_f([0, \infty)) = \sum_{k=0}^{N} x_k = \sum_{k=0}^{N} (U\Pi^{1/2}f)_k^2 = \sum_{k=0}^{N} \pi_k f_k^2 = \|f\|^2. \tag{3.27}$$

This explains the reason that we use symmetric matrix $A = \Pi^{1/2}Q\Pi^{-1/2}$ instead of ΠQ, which is also symmetric. Indeed, if we use the latter, then $F_f([0, \infty)) = \sum_k f_k^2$. This is meaningless in infinite space.

e) Next, since $U\Pi^{1/2}Q^n\Pi^{-1/2}U^* = \text{diag}\,((-\lambda_i)^n)$, by using the same argument as in the proof of (3.26), we have

$$(\Omega^n f, f) = \int_0^\infty (-\lambda)^n F_f(d\lambda).$$

From d), this holds also for $n = 0$. Combining this with the previous discussion (in the present case, $\Omega = Q$), we obtain

$$(P(t)f, f) = \int_0^\infty e^{-\lambda t} F_f(d\lambda). \tag{3.28}$$

f) Finally, from (3.27) we have $F_f([0, \infty)) = \|f\|^2 < \infty$. Inspired by (3.28), it is natural to define the measure-valued bi-linear form $(E_\lambda f, g)$ as follows.

$$d(E_\lambda f, f) := F_f(d\lambda), \quad \text{(a measure on } [0, \infty))$$

$$(E_\lambda f, g) := \frac{1}{4}\big[(E_\lambda(f + g), f + g) - (E_\lambda(f - g), f - g)\big].$$

(a signal measure)

Briefly speaking, for every fixed f, $(E_\lambda f, f)$ is a measure and for each set $C \in \mathscr{B}([0, \infty))$, $\int_C d(E_\lambda f, g)$ is bi-linear for f and g. We call $\{E_\lambda : \lambda \geqslant 0\}$ the **spectrum family** of operator Ω on $L^2(\pi)$. So, we have obtained (3.21). We should point out that, in the infinite space, it may happen that $\int_0^\infty \lambda d(E_\lambda f, f) = \infty$ for some $f \in L^2(\pi)$. Therefore, we have to deal carefully with the domain of operator L. Nevertheless, the bounded operator is still essential, since the unbounded operator can be approximated by the bounded ones in some way. However, the previous theorem is valid for the unbounded operator.

3.4 Supplements and Exercises

(1) Assume that the reversible Markov chain $P(t) = (p_{ij}(t))$ has stationary distribution π. Then $\lim_{t\to\infty} p_{ij}(t) = \pi_j$.

(2) Fix $T > 0$, define $Y_t = X_{T-t}$. Then

    (a) $(Y_t)_{0\leqslant t\leqslant T}$ is a Markov chain;

    (b) $(Y_t)_{0\leqslant t\leqslant T}$ has homogeneous transition probability matrix iff $(X_t)_{t\geqslant 0}$ is further stationary, that is the distribution of X_t $m_i = \mathbb{P}[X_t = i]$ is independent of t. In this case, the transition probability matrix of $(Y_t)_{0\leqslant t\leqslant T}$ is given by $p_{ij}^*(t) = m_j p_{ji}(t)/m_i$.

(3) Prove that the birth-death process is symmetrizable, and write out its symmetric measure. From this, give the necessary and sufficient condition for the positive recurrence.

(4) Use the fact that the forward equation and backward equation have the same nonnegative minimal solution, proving Theorem 3.4.

(5) Prove $(U\Pi^{1/2}f)(0) = \sum_i \pi_i f_i$ in (3.26).

(6) Let λ_1 be the spectral gap. Then

(a) $(-\Omega f, f) \geqslant \lambda_1 \left(\|f\|^2 - \left(\sum_i \pi_i f_i \right)^2 \right)$;

(b) $(P(t)f, f) - \left(\sum_i \pi_i f_i \right)^2 \leqslant e^{-\lambda_1 t} \left(\|f\|^2 - \left(\sum_i \pi_i f_i \right)^2 \right)$.

(7) Use Theorem 3.9 to give a "good" estimate for each of the following birth-death processes.

(a) $a_i = a$, $b_i = b$?

(b) $a_i = a$, $b_i = b/(i+1)$?

(c) $a_i = \delta i$, $b_i = \alpha i + \beta i$, where $\delta > \beta$.

(8) Assume that $P(t)$ has stationary distribution π. If

$$\lim_{t \to \infty} \sup_i \sum_j |p_{ij}(t) - \pi_j| = 0,$$

then we call $P(t)$ **strongly ergodic**. Prove that if $P(t)$ is strongly ergodic, then there exist $C < \infty$ and $\epsilon > 0$ such that

$$\sup_i \sum_j |p_{ij}(t) - \pi_j| \leqslant C e^{-\epsilon t}.$$

Consider the following arguments.

(a) Define $\phi(t) = \dfrac{1}{2} \sup_i \sum_j |p_{ij}(t) - \pi_j|$, then

$$\phi(t) \leqslant 1, \phi(t+s) \leqslant \phi(t)\phi(s).$$

(b) Define $T = \inf \{ t \geqslant 0 : \phi(t) \leqslant e^{-1} \}$, then

$$\phi(t) \leqslant \phi([t/T]T) \leqslant \phi(T)^{[t/T]}.$$

(9) Assume $(p_{ij}(t))$ is reversible with respect to π. For $H \subset E$, let $\tilde{\pi}_i = \pi_i/(1 - \pi(H))$ for each $i \notin H$ and $\tilde{\pi}(H) = 0$. Then

$$(P(t)f, g)_{\tilde{\pi}} = (f, P(t)g)_{\tilde{\pi}}.$$

(10) Prove that (3.18) is equivalent to (3.20) while (3.19) is equivalent to that $(\Omega f, g) = (f, \Omega g)$ for every $f, g \in \mathscr{D}(\Omega)$.

(11) If $\text{gap}(D) > 0$, then Markov chain is exponentially ergodic.

(12) For a regular Q-process $Q = (q_{ij})$, having reversible measure $\pi = (\pi_i)$, it holds that

$$\text{gap}(D) \leqslant \inf \left\{ \frac{1}{\pi(K)\pi(K^c)} \sum_{i \in K, j \notin K} \pi_i q_{ij} : 0 < \pi(K) < 1 \right\},$$

where $\pi(K) = \sum_{i \in K} \pi_i$ for a subset K.

(13) (a) We call $\phi : [0, \infty) \to [0, \infty)$ is a **superadditive function**, if for every $s, t > 0$, $\phi(s+t) \geqslant \phi(s) + \phi(t)$. Prove the limit $\sigma = \lim_{t \downarrow 0} \phi(t)/t$ exists and $\sigma = \inf_{t > 0} \phi(t)/t$.

(b) Let
$$\sigma(t) = -\sup \{\log \|P(t)f\| : \pi(f) = 0, \ \|f\| = 1\},$$
then $\sigma(t)$ is a superadditive function, so that limit $\sigma = \lim_{t \downarrow 0} \sigma(t)/t$ exists.

Hint: Use the Chapman-Kolmogorov equation.

(14) Assume a continuous-time Markov chain $P^{(i)}(t)$ on denumerable space E_i have Q-matrix $Q^{(i)}$ for $i = 1, 2$. Let the process $P(t) = (p_{(ij)(kl)}(t))$ on product space $E_1 \times E_2$ satisfies $p_{(ij)(kl)}(t) = p_{ik}^{(1)}(t) p_{jl}^{(2)}(t)$.

(a) Prove $P(t)$ is a Markov chain, and give its Q-matrix.

(b) Prove $\mathrm{gap}(Q) = \mathrm{gap}(Q^{(1)}) \bigwedge \mathrm{gap}(Q^{(2)})$.

(c) How about the corresponding n-fold product space?

(15) Prove
$$\lambda_0 \leqslant \lambda_1 \leqslant \lambda_0/\pi_0.$$
Hint: Use the fact that $\mathrm{Var}(f) = \pi(f^2) - \pi(f)^2 = \inf_{c \in \mathbb{R}} \pi((f - c)^2)$.

(16) Let (m_i) and (n_i) be two nonnegative sequences, satisfying
$$c := \sup_{i > 0} \sum_{j=0}^{i-1} n_j \sum_{j=i}^{\infty} m_j < \infty.$$
Prove that for every $\gamma \in (0, 1)$, we have
$$\sum_{j=i}^{\infty} \varphi_j^{\gamma} m_j \leqslant c(1 - \gamma)^{-1} \varphi_i^{\gamma - 1},$$
where $\varphi_k - \sum_{j=0}^{k-1} n_j$.

(17) (Hardy's inequality) Let
$$\delta = \sup_{n \geqslant 1} \sum_{i=0}^{n-1} \frac{1}{\mu_j b_j} \sum_{i=n}^{\infty} \mu_i.$$
Prove $(4\delta)^{-1} \leqslant \lambda_0 \leqslant \delta^{-1}$.

Hint: In the previous exercise, by letting $\gamma = 1/2, m_i = \mu_i, n_i = (\mu_i b_i)^{-1}$, we have the lower estimate. On the other hand, in the definition of λ_0, by letting
$$f_i = \sum_{j=0}^{(i-1) \wedge (n-1)} \frac{1}{\mu_j b_j},$$
we have $\lambda_0 \leqslant \delta^{-1}$, where λ_0 is defined above Theorem 3.14.

(18) Assume that μ, ν are two signed measures on denumerable space E. Prove that

(a) $\|\mu \times \nu\|_{\mathrm{Var}} \leqslant \|\mu\|_{\mathrm{Var}} + \|\nu\|_{\mathrm{Var}}$; (The definition of $\|\cdot\|_{\mathrm{Var}}$ can be found in Exercise 32 in Chapter 2.) From this, prove that

(b) (Following Exercises 8 and 14) $P(t)$ is strongly ergodic iff so is $P^{(i)}(t)(i = 1, 2)$.

(19) Let

$$
Q = \begin{pmatrix} -3 & 1 & 2 \\ 2 & -4 & 2 \\ 2 & 1 & -3 \end{pmatrix}.
$$

Prove its transition probability matrix is

$$
P(t) = \frac{1}{5} \begin{pmatrix} 2 + 3e^{-5t} & 1 - e^{-5t} & 2 - 2e^{-5t} \\ 2 - 2e^{-5t} & 1 + 4e^{-5t} & 2 - 2e^{-5t} \\ 2 - 2e^{-5t} & 1 - e^{-5t} & 2 + 3e^{-5t} \end{pmatrix}.
$$

So the limiting distribution of $P(t)$ is $\pi = (2/5, 1/5, 2/5)$, and

$$
\sup_i \sum_j |p_{ij}(t) - \pi_j| = \frac{8}{5} e^{-5t}.
$$

(20) Assume

$$
Q = \begin{pmatrix} -1 & 1 & 0 \\ 0 & -1 & 1 \\ 1 & 0 & -1 \end{pmatrix}.
$$

Prove its transition probability matrix $P(t)$ is

$$
\frac{1}{3} \begin{pmatrix} 1 & 1 & 1 \\ 1 & 1 & 1 \\ 1 & 1 & 1 \end{pmatrix} + \frac{2}{3} e^{-3t/2} R(t),
$$

where $R(t)$ is

$$
\begin{pmatrix} \cos(\frac{\sqrt{3}t}{2}) & \cos(\frac{\sqrt{3}t}{2} - \frac{2\pi}{3}) & \cos(\frac{\sqrt{3}t}{2} + \frac{2\pi}{3}) \\ \cos(\frac{\sqrt{3}t}{2} + \frac{2\pi}{3}) & \cos(\frac{\sqrt{3}t}{2}) & \cos(\frac{\sqrt{3}t}{2} - \frac{2\pi}{3}) \\ \cos(\frac{\sqrt{3}t}{2} - \frac{2\pi}{3}) & \cos(\frac{\sqrt{3}t}{2} + \frac{2\pi}{3}) & \cos(\frac{\sqrt{3}t}{2}) \end{pmatrix}.
$$

So the limiting distribution of $P(t)$ is $\pi = (1/3, 1/3, 1/3)$, and

$$
\frac{2}{3} e^{-3t/2} \leqslant \sup_i \sum_j |p_{ij}(t) - \pi_j| \leqslant \frac{4}{3} e^{-3t/2}.
$$

Chapter 4

General Markov Processes

4.1 Markov Property and Its Equivalence

In this chapter, we study the Markov processes on general state spaces.

Definition 4.1. A stochastic process $(X_t)_{t \geqslant 0}$ is called a **Markov process** on probability space $(\Omega, \mathscr{F}, \mathbb{P})$, taking values in a measurable space $(E, \mathscr{E})$, if the following **Markov property** is fulfilled: for every $n \geqslant 2, 0 \leqslant t_1 < t_2 < \cdots < t_n < u$ and $A_u \in \mathscr{E}$,

$$\mathbb{P}[X_u \in A_u | X_{t_1}, \cdots, X_{t_n}] = \mathbb{P}[X_u \in A_u | X_{t_n}]. \tag{4.1}$$

Recall by the definition of conditional probability, for $\mathscr{G}_1 \subset \mathscr{G}_2 \subset \mathscr{F}$, we have $\mathbb{E}[f|\mathscr{G}_i] \in \mathscr{G}_i$ ($i = 1, 2$) and

$$\mathbb{E}[f|\mathscr{G}_1] = \mathbb{E}[f|\mathscr{G}_2] \Longleftrightarrow \int_\Lambda \mathbb{E}[f|\mathscr{G}_1] \mathrm{d}\mathbb{P} = \int_\Lambda f \mathrm{d}\mathbb{P}, \quad \Lambda \in \mathscr{G}_2.$$

The purpose of this section is to deduce the equivalent statements of (4.1). The main tool is the monotone class theorem in the measure theory.

Let

$$\mathscr{C}_t = \{ [X_{t_1} \in A_{t_1}, \cdots, X_{t_n} \in A_{t_n}] : A_{t_k} \in \mathscr{E}, 0 \leqslant t_1 < \ldots < t_n \leqslant t, n \geqslant 1 \},$$

it is a π-class which generates σ-algebra

$$\mathscr{F}_t := \sigma(\mathscr{C}_t) = \sigma\{X_s : s \leqslant t\}.$$

It follows from (4.1) that for each $B \in \mathscr{C}_t$, say,

$$B = [X_{t_1} \in A_{t_1}, \cdots, X_{t_n} \in A_{t_n}],$$

we have

$$\int_B \mathbb{P}[X_u \in A_u | X_t] \mathrm{d}\mathbb{P} = \int_B \mathbb{P}[X_u \in A_u | X_{t_1}, \cdots, X_{t_n}, X_t] \mathrm{d}\mathbb{P}$$

$$= \mathbb{E}\big[\mathbb{E}\big[I_{[X_u \in A_u]} | X_{t_1}, \cdots, X_{t_n}, X_t\big] I_B\big]$$

$$= \mathbb{E}\big[\mathbb{E}\big[I_{[X_u \in A_u]} I_B | X_{t_1}, \cdots, X_{t_n}, X_t\big]\big]$$

$$= \mathbb{E}[I_{[X_u \in A_u]} I_B].$$

Thus by the monotone class theorem for sets,

$$\mathbb{P}[X_u \in A_u | \mathscr{F}_t] = \mathbb{P}[X_u \in A_u | X_t], \qquad u > t. \tag{4.2}$$

Then by using the monotone class theorem, we can replace the indicator function $I_{[X_u \in A_u]}$ by any $\sigma(X_u)$-measurable function:

$$\mathbb{E}[\xi | \mathscr{F}_t] = \mathbb{E}[\xi | X_t], \qquad \xi \in \sigma(X_u), \ \mathbb{E}|\xi| < \infty. \tag{4.3}$$

In equation (4.3), we consider only one epoch in the future. We are now going to replace it by finitely many epochs in the future.

For any finite many $t < u_1 < \cdots < u_m$ and $A_k \in \mathscr{E}$,

$$\mathbb{P}[X_{u_j} \in A_j, j = 1, 2, \cdots, m | \mathscr{F}_t] = \mathbb{P}[X_{u_j} \in A_j, j = 1, 2, \cdots, m | X_t]. \tag{4.4}$$

Now, we can mimic the proofs of (4.1) $\implies$ (4.2) $\implies$ (4.3), where we add the epochs in the past step by step. Here we add the epochs in the future step by step. Let

$$\mathscr{C}^t = \{[X_{u_1} \in A_1, X_{u_2} \in A_2, \ldots, X_{u_m} \in A_m] :$$
$$u_m > u_{m-1} > \ldots > u_1 \geqslant t, \ A_k \in \mathscr{E}, \ m \geqslant 1\},$$
$$\mathscr{F}^t = \sigma(\mathscr{C}^t) = \sigma\{X_u : u \geqslant t\}.$$

Then the monotone class theorem implies

$$\mathbb{P}[B | \mathscr{F}_t] = \mathbb{P}[B | X_t], \qquad B \in \mathscr{F}^t, \tag{4.5}$$

and furthermore

$$\mathbb{E}[\xi | \mathscr{F}_t] = \mathbb{E}[\xi | X_t], \qquad \xi \in \mathscr{F}^t, \ \mathbb{E}|\xi| < \infty. \tag{4.6}$$

In the sequel, we need the following representation theorem for measurable functions on a product space.

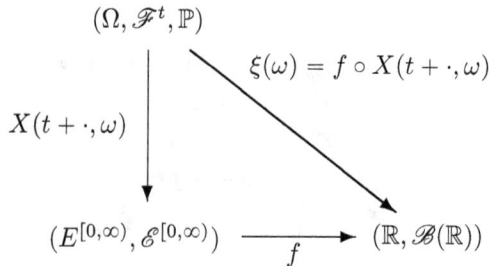

That is, for any measurable mapping ξ from $(\Omega, \mathscr{F}^t, \mathbb{P})$ to $(\mathbb{R}, \mathscr{B}(\mathbb{R}))$, there exists a measurable mapping f from $(E^{[0,\infty)}, \mathscr{E}^{[0,\infty)})$ to $(\mathbb{R}, \mathscr{B}(\mathbb{R}))$, such that ξ is the composition of f and $X(t + \cdot)$. Thus, it follows from (4.6) that for every $f \in \mathscr{E}^{[0,\infty)}$,

$$\mathbb{E}[f \circ X(t + \cdot) | \mathscr{F}_t] = \mathbb{E}[f \circ X(t + \cdot) | X_t], \qquad \mathbb{E}|f \circ X(t + \cdot)| < \infty. \tag{4.7}$$

Now let $\xi \in \mathscr{F}^t$, $\eta \in \mathscr{F}_t$ such that ξ, η and $\xi\eta$ are integrable. Then

$$
\begin{aligned}
\mathbb{E}[\xi\eta|X_t] &= \mathbb{E}[\mathbb{E}[\xi\eta|\mathscr{F}_t]|X_t] = \mathbb{E}[\eta\mathbb{E}[\xi|\mathscr{F}_t]|X_t] \\
&= \mathbb{E}[\eta\mathbb{E}[\xi|X_t]|X_t] \qquad \text{(by (4.7))} \\
&= \mathbb{E}[\xi|X_t]\mathbb{E}[\eta|X_t].
\end{aligned} \tag{4.8}
$$

The advantage of this equation is that the "past" and the "future" are in symmetric positions.

Now we can state the main theorem in this section.

Theorem 4.2. Properties (4.1)–(4.8) are all equivalent.

Proof We have proved that

$$(4.1) \Longrightarrow (4.2) \Longrightarrow (4.3)$$

and

$$(4.4) \Longrightarrow (4.5) \Longrightarrow (4.6) \Longrightarrow (4.7) \Longrightarrow (4.8).$$

So we need only prove that $(4.3) \Longrightarrow (4.4)$ and $(4.8) \Longrightarrow (4.1)$.

$(4.3) \Longrightarrow (4.4)$.   We use the induction method. When $m = 1$, (4.4) is just (4.3). Let $B_1 = [X_{u_1} \in A_1]$, $B_2 = [X_{u_2} \in A_2, \cdots, X_{u_m} \in A_m]$. Then

$$
\begin{aligned}
\mathbb{E}[I_{B_1 B_2}|\mathscr{F}_t] &= \mathbb{E}[\mathbb{E}[I_{B_1 B_2}|\mathscr{F}_{u_1}]|\mathscr{F}_t] = \mathbb{E}[I_{B_1}\mathbb{P}[B_2|\mathscr{F}_{u_1}]|\mathscr{F}_t] \\
&= \mathbb{E}[I_{B_1}\mathbb{P}[B_2|X_{u_1}]|\mathscr{F}_t] \quad \text{(by the induction assumption)} \\
&= \mathbb{E}[I_{B_1}\mathbb{P}[B_2|X_{u_1}]|X_t]. \quad \text{(by (4.3))}
\end{aligned}
$$

Similarly,

$$
\begin{aligned}
\mathbb{E}[I_{B_1 B_2}|X_t] &= \mathbb{E}[\mathbb{E}[I_{B_1 B_2}|\mathscr{F}_{u_1}]|X_t] = \mathbb{E}[I_{B_1}\mathbb{P}[B_2|\mathscr{F}_{u_1}]|X_t] \\
&= \mathbb{E}[I_{B_1}\mathbb{P}[B_2|X_{u_1}]|X_t].
\end{aligned}
$$

$(4.8) \Longrightarrow (4.1)$ It follows from (4.8) that $\mathbb{E}[\xi\eta|X_t] = \mathbb{E}[\eta\mathbb{E}[\xi|X_t]|X_t]$. Thus, for every $A \in \mathscr{E}$,

$$\int_{[X_t \in A]} \xi\eta \, d\mathbb{P} = \int_{[X_t \in A]} \eta E[\xi|X_t] d\mathbb{P}.$$

In particular, if we set $\xi = I_{[X_u \in A_u]}$, $\eta = I_{[X_{t_1} \in A_1, \cdots, X_{t_n} \in A_n]}$, then

$$\mathbb{P}[X_{t_1} \in A_1, \cdots, X_{t_n} \in A_n, X_t \in A, X_u \in A_u]$$

$$= \int_{[X_t \in A]} \xi\eta \, d\mathbb{P}$$

$$= \int_{[X_{t_1} \in A_1, \cdots, X_{t_n} \in A_n, X_t \in A]} \mathbb{P}[X_u \in A_u|X_t] d\mathbb{P}.$$

This is just (4.1). $\square$

If we add some conditions on the state space $(E, \mathscr{E})$, ensuring the existence of regular conditional probability, then the Markov process should have a transition probability function. In other words, $\mathbb{P}[X_t \in A | X_s]$ has a modification $p(s, x; t, A)$, which is a transition probability function. In the following, assume that $\mathscr{E}$ contains all singleton $\{x\}$, $x \in E$.

Definition 4.3. We call the 4-variate function $p(s, x; t, A)(t \geqslant s \geqslant 0, x \in E, A \in \mathscr{E})$ a **transition probability function**, if

  (1) for given s, x, t, it is a probability measure of A on $\mathscr{E}$;
  (2) for given s, t, A, it is a measurable function of x in E;
  (3) $p(s, x; s, \{x\}) = 1$ for every $s \geqslant 0$ and $x \in E$;
  (4) it satisfies the Chapman-Kolmogorov equation: for any $s \leqslant t \leqslant u$,

$$p(s, x; u, A) = \int_E p(s, x; t, \mathrm{d}y)\, p(t, y; u, A).$$

It is called **time-homogeneous**, if $p(s, x; t, A) = p(t - s, x, A)$ for every $t \geqslant s, x \in E, A \in \mathscr{E}$.

Given a Markov process, denote $P_t(A) = \mathbb{P}[X_t \in A]$. Then by the Markov property and the induction method, its finite-dimensional distributions are given as follows.

For $0 = t_0 < t_1 < \cdots < t_n$ and $A_k \in \mathscr{E}(1 \leqslant k \leqslant n)$,

$$\mathbb{P}[X_{t_k} \in A_k, k = 1, \cdots, n] = \int_{A_0} P_{t_0}(\mathrm{d}x_0) \int_{A_1} p(t_0, x_0; t_1, \mathrm{d}x_1)$$

$$\times \int_{A_2} p(t_1, x_1; t_2, \mathrm{d}x_2) \cdots \int_{A_n} p(t_{n-1}, x_{n-1}; t_n, \mathrm{d}x_n).$$

$$(4.9)$$

Equivalently (use the monotone class theorem to prove it), for every n-variate nonnegative function f, we have

$$\mathbb{E}[f(X_{t_0}, X_{t_1}, \cdots, X_{t_n})] = \int P_{t_0}(\mathrm{d}x_0) \int p(t_0, x_0; t_1, \mathrm{d}x_1) \int p(t_1, x_1; t_2, \mathrm{d}x_2)$$

$$\cdots \int p(t_{n-1}, x_{n-1}; t_n, \mathrm{d}x_n) f(x_0, x_1, \cdots, x_n).$$

$$(4.10)$$

Conversely, for any initial distribution P_0 and transition probability function $p(s, x; t, A)$, if the finite-dimensional distributions of a stochastic process satisfy (4.9), then it must be Markovian. In other words, the Markov process is determined completely by the initial distribution and the transition probability function.

Next, we present some examples of Markov processes on the real line $\mathbb{R}$.

Example 4.4 (Standard Brownian motion). Its transition probability density is the **Gaussian kernel**:

$$\varphi(t, x, y) = \frac{1}{\sqrt{2\pi t}} \exp\left[-\frac{(y-x)^2}{2t} \right].$$

It satisfies the heat equation $\partial\varphi/\partial t = \frac{1}{2}\partial^2\varphi/\partial x^2$. That is, it corresponds to Laplacian operator $\frac{1}{2}\Delta$.

Here the differential operator is the analog of Q-matrix in the discrete space.

Example 4.5 (Brownian motion with drift). Its transition probability density is:

$$p(t, x, y) = \varphi(\sigma^2 t, x + \mu t, y).$$

The corresponding backward equation $\partial p/\partial t = \frac{1}{2}\sigma^2\partial^2 p/\partial x^2 + \mu\partial p/\partial x$.

Example 4.6 (Ornstein-Uhlenbeck process, or O.U. process). Its transition probability density is:

$$p(t, x, y) = \varphi\left(\sigma^2\left(1 - e^{-2\alpha t}\right)/(2\alpha), xe^{-\alpha t}, y\right).$$

The corresponding backward equation is $\partial p/\partial t = \frac{1}{2}\sigma^2\partial^2 p/\partial x^2 - \alpha x\partial p/\partial x$.

All these processes are symmetrizable, and O.U. process is reversible with respect to the normal distribution.

The direct generalization of the continuous-time Markov chain is the following

Example 4.7 (General jump process). Let $(E, \mathcal{E})$ be a measurable space such that $\mathcal{E}$ contains all singletons. Instead of Q-matrix, we use q-pair $(q(x), q(x, dy))$, where $q(x, dy)$ is a nonnegative measurable kernel and $q(x)$ is a measurable function, such that $q(x, E) = q(x, E\backslash\{x\}) \leqslant q(x) \leqslant \infty$. The total stability and conservation mean that $q(x) = q(x, E) < \infty$ for each $x \in E$. The so-called **jump process**, is a transition probability $p(t, x, dy)$, satisfying **jump condition**: $\lim_{t\downarrow 0} p(t, x, \{x\}) = 1$ for every $x \in E$. Thus, as in the discrete state space, we can find a theory for the general jump process. Refer to [5; 10].

The main theorem in this section (Theorem 4.2) is also suitable for the discrete-time Markov process $(X_n)_{n\geqslant 0}$. As an example, we have

Example 4.8 (Economic model). Denote by X_n the vector of product in the nth year. Then $(X_n)_{n\geqslant 0}$ is a Markov process valued in $\mathbb{R}^d$.

4.2   Strong Markov Property

In this section, we study the core problem in the theory of Markov process: strong Markov property.

Let $(X_t)_{t \geqslant 0}$ be a stochastic process defined on a probability space $(\Omega, \mathscr{F}, \mathbb{P})$, taking values in the measurable space $(E, \mathscr{E})$. Denote $\mathscr{F}_t := \sigma(X_s, s \leqslant t)$. It is called a **Markov process**, if the following **Markov property** holds

$$\mathbb{E}[f(X_{s+t})|\mathscr{F}_s] = \mathbb{E}[f(X_{s+t})|X_s] =: \mathbb{E}_{s,X_s}[f(X_{s+t})], \quad t, \ s \geqslant 0. \quad (4.11)$$

In other words, conditional on "present" (a fixed epoch s), the "future" depends only on "present" and is independent of "past". Now suppose we are going to observe the behavior of the process after the first time arriving at some state or set, regarding this first arriving time T as "present", then the question is whether the future behavior of the process depends or not on "present" and its position X_T, but is it independent of the past? In other words, do we have

$$\mathbb{E}[f(X_{T+t})|\mathscr{F}_T] = \mathbb{E}[f(X_{T+t})|X_T] = \mathbb{E}_{T,X_T}[f(X_{T+t})], \quad t \geqslant 0? \quad (4.12)$$

Notice that T is random. Of course, (4.12) is much more extensive than (4.11). We call (4.12) the **strong Markov property**. Historically, people thought (4.12) would be a trivial extension of (4.11). Until 1950's, some counterexamples were discovered, which attracted much more attention to the strong Markov property. There are 2 questions. Firstly, can (4.12) hold for what kind T? Secondly, how to define the σ-algebra $\mathscr{F}_T$? The purpose of this section is to answer these 2 questions. In the following, we introduce a method, which can be served as a sample of doing mathematical research.

Now, let us consider the basic case: the process has time-homogeneous transition probability function $p(t, x, dy)$. Namely,

$$\mathbb{E}_{s,X_s}[f(X_{s+t})] = \mathbb{E}_{0,X_s}[f(X_t)] =: \mathbb{E}_{X_s}[f(X_t)] = \int p(t, X_s, \mathrm{d}y) f(y).$$

Therefore Markov property (4.11) becomes

$$\mathbb{E}[f(X_{s+t})|\mathscr{F}_s] = \mathbb{E}_{X_s}[f(X_t)],$$

while (4.12) becomes

$$\mathbb{E}[f(X_{T+t})|\mathscr{F}_T] = \mathbb{E}_{X_T}[f(X_t)]. \quad (4.13)$$

Let us begin with a simple "random time" T:

$$T = \sum_{i=1}^{n} a_i I_{A_i}, \qquad \sum_{i=1}^{n} A_i = \Omega, \quad a_i \geqslant 0,$$

where a_i's are different and A_i's are disjoint. Thus, (4.13) holds iff

$$\int_A f(X_{T+t})\mathrm{d}\mathbb{P} = \int_A \mathbb{E}_{X_T} f(X_t)\mathrm{d}\mathbb{P}, \qquad A \in \mathscr{F}_T.$$

But by Markov property,

$$\int_A \mathbb{E}_{X_T} f(X_t)\mathrm{d}\mathbb{P} = \sum_{i=1}^n \int_{AA_i} \mathbb{E}_{X_T} f(X_t)\mathrm{d}\mathbb{P} = \sum_{i=1}^n \int_{AA_i} \mathbb{E}_{X_{a_i}} f(X_t)\mathrm{d}\mathbb{P}$$

$$= \sum_{i=1}^n \int_{AA_i} \mathbb{E}[f(X_{a_i+t})|\mathscr{F}_{a_i}]\mathrm{d}\mathbb{P}. \qquad (4.14)$$

In order to put AA_i into the conditional expectation we need $AA_i \in \mathscr{F}_{a_i}$, by using the smooth property of conditional expectation. If this was the case, then we would obtain

$$\text{the right-hand side of (4.14)} = \sum_{i=1}^n \int \mathbb{E}[I_{AA_i} f(X_{a_i+t})|\mathscr{F}_{a_i}]\mathrm{d}\mathbb{P}$$

$$= \sum_{i=1}^n \mathbb{E}[I_{AA_i} f(X_{a_i+t})]$$

$$= \mathbb{E}\left[I_A \sum_{i=1}^n I_{A_i} f(X_{a_i+t})\right] = \mathbb{E}[I_A f(X_{T+t})]$$

$$= \int_A f(X_{T+t})\mathrm{d}\mathbb{P}.$$

Combining this with (4.14), we obtain (4.13). Thus, in order that (4.13) holds, we need only

$$A \in \mathscr{F}_T \implies AA_i = A \cap [T = a_i] \in \mathscr{F}_{a_i}, \quad \forall i. \qquad (4.15)$$

As $\mathscr{F}_T$ is a σ-algebra, Ω must be in $\mathscr{F}_T$. By taking $A = \Omega$ in (4.15), we get a condition for T:

$$[T = a_i] \in \mathscr{F}_{a_i}.$$

Since T is discrete, this condition is equivalent to $[T \leqslant a_i] \in \mathscr{F}_{a_i}$. For general T, since $[T = a_i]$ is usually a null probability set, we will use the latter condition $[T \leqslant t] \in \mathscr{F}_t$. This argument leads to the following definition.

Definition 4.9. Assume $T : \Omega \to [0, +\infty]$ is a random variable. If $[T \leqslant t] \in \mathscr{F}_t$ for every $t \geqslant 0$, then T is called a **stopping time** (or **optional time**, or **random variable independent of the future**).

Notice that the stopping time may take value $+\infty$. Before moving further, we study the properties of the stopping times.

Theorem 4.10.
(1) The constant time is a stopping time.
(2) If T is a stopping time and $t \geqslant 0$, then so is $T + t$.
(3) If T_1 and T_2 are stopping times, then so is $T_1 + T_2$.
(4) If T_1 and T_2 are stopping times, then so are $T_1 \vee T_2 := \max\{T_1, T_2\}$ and $T_1 \wedge T_2 := \min\{T_1, T_2\}$.
(5) If T_n are stopping times and $T_n \uparrow T$, then so is T.
(6) Assume T is a stopping time. Let

$$T_n = \sum_{k=0}^{\infty} \frac{k+1}{2^n} I_{[k/2^n < T \leqslant (k+1)/2^n]} + \infty I_{[T=\infty]}.$$

Then T_n are stopping times, and $T_n \downarrow T$.

Proof We prove only (6). For each $t \in [0, \infty)$, we have

$$[T_n \leqslant t] = \sum_{k:\,(k+1)/2^n \leqslant t} \left[\frac{k}{2^n} < T \leqslant \frac{k+1}{2^n} \right].$$

But

$$\left[\frac{k}{2^n} < T \leqslant \frac{k+1}{2^n} \right] = \left[T \leqslant \frac{k+1}{2^n} \right] \bigcap \left[T \leqslant \frac{k}{2^n} \right]^c \in \mathscr{F}_{\frac{k+1}{2^n}} \subset \mathscr{F}_t.$$

Thus $[T_n \leqslant t] \in \mathscr{F}_t$. $\square$

Remark 4.11. To approximate a given random variable T, one often adopts the following sequence of simple random variables:

$$\tilde{T}_n = \sum_{k=0}^{n2^n - 1} \frac{k}{2^n} I_{[k/2^n \leqslant T < (k+1)/2^n]} + n I_{[T=\infty]}.$$

Then we have $\tilde{T}_n \uparrow T$. But

$$[\tilde{T}_n \leqslant t] = \sum_{k/2^n \leqslant t} \left[\frac{k}{2^n} \leqslant T < \frac{k+1}{2^n} \right] = \sum_{k/2^n \leqslant t} \left[T < \frac{k+1}{2^n} \right] \bigcap \left[T < \frac{k}{2^n} \right]^c.$$

The sets on the right-hand side may not belong to $\mathscr{F}_t$. This shows for a general stopping time, it may not be approximated from the left.

Now we consider the second problem: how to define $\mathscr{F}_T$? Indeed, as we have seen in (4.15), a set A in $\mathscr{F}_T$ must satisfies: $A \cap [T = a_i] \in \mathscr{F}_{a_i}$ for every i. This is equivalent to that $A \cap [T \leqslant a_i] \in \mathscr{F}_{a_i}$ for all i. For the general case (not necessary discrete), we should define

Definition 4.12. We call $\mathscr{F}_T := \{A \in \mathscr{F} : A \cap [T \leqslant t] \in \mathscr{F}_t \text{ for every } t \geqslant 0\}$ the **pre-T σ-algebra**.

The following property shows that $\mathscr{F}_T$ is not too small, and it meets one of our fundamental requirements.

Lemma 4.13. $T \in \mathscr{F}_T$.

Proof For every s, $t \geqslant 0$, we have

$$[T \leqslant s] \cap [T \leqslant t] = [T \leqslant t \wedge s] \in \mathscr{F}_{t \wedge s} \subset \mathscr{F}_t.$$

Thus $[T \leqslant s] \in \mathscr{F}_T$ for every s. $\square$

We have thus proved the strong Markov property (4.13) in the case of simple stopping times. For general stopping times, naturally we want to approximate them by using simple ones.

First look at the left-hand side in (4.13). We hope

$$\mathbb{E}[f(X_{T_n + t}) | \mathscr{F}_{T_n}] \to \mathbb{E}[f(X_{T + t}) | \mathscr{F}_T] \qquad \text{as } n \to \infty.$$

For the time being, forget the condition "$\mathscr{F}_T$" and fix t. It is easy to see that this requires some continuity on X_t and f. At least it is hopeful when both are continuous. Can we weaken this condition? As for f, it is enough to assume it is bounded and continuous. Since if (4.13) holds for this f, then by applying the monotone class theorem, (4.13) holds for every bounded measurable functions. As for X_t, when $T_n \downarrow T$, it is enough to assume that X_t is right-continuous, and then we have $X_{T_n + t} \to X_{T + t}$.

Next look at the right-hand side in (4.13). Fix t. What we require is

$$\mathbb{E}_{X_{T_n}}[f(X_t)] \to \mathbb{E}_{X_T}[f(X_t)], \qquad \text{as } n \to \infty, \quad f \in C_b,$$

where C_b denotes the total of bounded continuous functions. This requires $\mathbb{E}_x f(X_t)$ is continuous in x. Notice

$$\mathbb{E}_x f(X_t) = \int f(y) p(t, x, \mathrm{d}y) =: P(t) f(x).$$

Thus the above condition can be rewritten as: $P(t)f(x)$ is continuous in x and bounded for each $f \in C_b$.

Definition 4.14. Assume $(E, \mathscr{E})$ is a measurable metric space. We call the transition function $p(t, x, A)$ a **Feller transition probability function**, if $P(t)f \in C_b$ for every $f \in C_b$ and every $t \geqslant 0$. The Markov process having this property is called **Feller process**.

Definition 4.15. We call the homogeneous Markov process (X_t) **strong Markov process**, if

(1) $X_T \in \mathscr{F}_T$ for every stopping time T;

(2) for every $f \in {}_b\mathscr{E} :=$ the total of bounded $\mathscr{E}$-measurable functions,

$$\mathbb{E}[f(X_{T+t}) | \mathscr{F}_T] = \mathbb{E}_{X_T}[f(X_t)] = P(t)f(X_T), \qquad t \geqslant 0, \quad (4.16)$$

where $X_T(\omega) = X_{T(\omega)}(\omega)$.

Notice that when $T = \infty$, X_T is not defined. Therefore, once X_T appears, it is understood as $X_T I_{[T<\infty]}$.

Now, we can state the main theorem of this section.

Theorem 4.16. The right-continuous Feller process is strong Markov process.

Proof a) Prove the measurability: $X_T \in \mathscr{F}_T$. Firstly, we prove conditional on $[T \leqslant t]$, $(T(\omega), \omega)$ is a measurable mapping from $(\Omega, \mathscr{F}_t)$ to $([0, t) \times \Omega, \mathscr{B}[0, t) \times \mathscr{F}_t)$. Indeed, for every $B \in \mathscr{B}[0, t)$ and $\Lambda \in \mathscr{F}_t$,

$$[(T(\omega), \omega)^{-1}(B \times \Lambda)] \cap [T \leqslant t] = [T \leqslant t] \cap T^{-1}(B) \cap \Lambda \in \mathscr{F}_t.$$

Secondly, in the following Lemma 4.18, we will prove: $X_t(\omega)$ is a measurable mapping from $([0, t) \times \Omega, \mathscr{B}[0, t) \times \mathscr{F}_t)$ to $(E, \mathscr{E})$. Thus, conditional on $[T \leqslant t]$, X_T is a composition mapping of $(T(\omega), \omega)$ and $X(t, \omega)$.

$$(\Omega, \mathscr{F}_t) \longrightarrow ([0, t) \times \Omega, \mathscr{B}[0, t) \times \mathscr{F}_t) \longrightarrow (E, \mathscr{E}),$$

which is surely measurable.

b) Prove (4.16). Let

$$T_n = \sum_{k=0}^{\infty} \frac{k+1}{2^n} I_{[k/2^n < T \leqslant (k+1)/2^n]} + \infty I_{[T=\infty]}.$$

Then T_n are stopping times by Lemma 4.10, and $T_n \downarrow T$. To prove (4.16), we fix $f \in C_b$. It suffices to prove for every $A \in \mathscr{F}_T$,

$$\int_A \mathbb{E}(f(X_{T+t}) | \mathscr{F}_T) d\mathbb{P} = \int_A P(t) f(X_T) d\mathbb{P}.$$

That is

$$\int_A f(X_{T+t}) d\mathbb{P} = \int_A P(t) f(X_T) d\mathbb{P}.$$

By the right-continuity of (X_t) and Feller property, we have

$$\int_A P(t) f(X_T) d\mathbb{P} = \lim_{n \to \infty} \int_A P(t) f(X_{T_n}) d\mathbb{P}$$

$$= \lim_{n \to \infty} \sum_{k=0}^{\infty} \int_{A \cap \{\frac{k}{2^n} < T \leqslant \frac{k+1}{2^n}\}} P(t) f\left(X_{\frac{k+1}{2^n}}\right) d\mathbb{P}$$

$$= \lim_{n \to \infty} \sum_{k=0}^{\infty} \int_{A \cap \{\frac{k}{2^n} < T \leqslant \frac{k+1}{2^n}\}} \mathbb{E}\left[f\left(X_{\frac{k+1}{2^n}+t}\right) \Big| \mathscr{F}_{\frac{k+1}{2^n}}\right] d\mathbb{P}$$

(by Markov property)

$$= \lim_{n \to \infty} \sum_{k=0}^{\infty} \int \mathbb{E}\left[I_{A \cap \{\frac{k}{2^n} < T \leqslant \frac{k+1}{2^n}\}} f\left(X_{\frac{k+1}{2^n}+t}\right) \Big| \mathscr{F}_{\frac{k+1}{2^n}}\right] d\mathbb{P}$$

$$\left(A \in \mathscr{F}_T \implies A \bigcap \left\{ \frac{k}{2^n} < T \leqslant \frac{k+1}{2^n} \right\} \in \mathscr{F}_{\frac{k+1}{2^n}} \right)$$

$$= \lim_{n \to \infty} \sum_{k=0}^{\infty} \int_A I_{A \cap \{ \frac{k}{2^n} < T \leqslant \frac{k+1}{2^n} \}} f\left(X_{\frac{k+1}{2^n}+t} \right) \mathrm{d}\mathbb{P}$$

$$= \lim_{n \to \infty} \sum_{k=0}^{\infty} \int_A I_{\{ \frac{k}{2^n} < T \leqslant \frac{k+1}{2^n} \}} f\left(X_{\frac{k+1}{2^n}+t} \right) \mathrm{d}\mathbb{P}$$

$$= \lim_{n \to \infty} \int_A f\left(X_{T_n+t} \right) \mathrm{d}\mathbb{P} = \int_A f(X_{T+t}) \mathrm{d}\mathbb{P}. \quad \square$$

Definition 4.17. We call (X_t) **adapted**, if $X_t \in \mathscr{F}_t$ for every $t \geqslant 0$. We call (X_t) **progressively measurable**, if $(X_s(\omega) : 0 \leqslant s \leqslant t, \omega \in \Omega)$ is a measurable mapping from $([0,t) \times \Omega, \mathscr{B}[0,t) \times \mathscr{F}_t)$ to $(E, \mathscr{E})$ for every $t \geqslant 0$.

Lemma 4.18. The right-continuous and adapted process is progressively measurable.

Proof It suffices to consider the following approximation:

$$X_s^{(n)} = X_0 I_{[s=0]} + \sum_{k=1}^{2^n} X_{kt/2^n} I_{[(k-1)t/2^n < s \leqslant kt/2^n]}. \quad \square$$

4.3 Appendix: Optimal Stopping Problem—The Secretary Problem

In this section, we show the importance of stopping time via a concrete problem.

A manager wants to hire a secretary from N girls. Every time he interviews only one girl. After the interview, he decides to hire her or not immediately. If a girl is refused, she has no more chance to come back. The manager will hire a girl according to her rank among the interviewed girls. Of course, he has no information on the girls waiting for interview. The problem is how should the manager proceed, to ensure he can choose the best one among these N girls with greatest probability.

According to the total scores of N girls, their ranks are $\{1, 2, \cdots, N\}$. Let Ω be all permutations $(\omega_1, \omega_2, \cdots, \omega_N)$ of $\{1, 2, \cdots, N\}$, and $\mathscr{F}$ be the total of all subsets of Ω. Let

$$\mathbb{P}[(\omega_1, \omega_2, \cdots, \omega_N)] = 1/N!.$$

Denote by

$$Y_n(\omega) = \#\{j : \omega_j \leqslant \omega_n\},$$

which means the relative rank of the nth girl after n interviews, and by $\mathscr{F}_n = \sigma\{Y_m : m \leqslant n\}$, which is the whole information until the nth interview. Thus, the problem is transferred into: seeking a stopping time T^* with respect to $(\mathscr{F}_n)$, such that $\mathbb{P}[\omega_{T*} = 1] = \max\{\mathbb{P}[\omega_T = 1] : T$ is a stopping time$\}$.

However, general stopping times are too complicated to deal with, so we restrict ourself to the following class of stopping times:
$$T_r = \min\{n \geqslant r : Y_n = 1\}, \quad 1 \leqslant r \leqslant N,$$
which means the epoch to find the best girl (relatively) in the rth interview and after. Our aim is to find r^* such that
$$\mathbb{P}[\omega_{T_{r*}} = 1] = \max_{1 \leqslant r \leqslant N} \mathbb{P}[\omega_{T_r} = 1].$$
Notice $Y_1, Y_2, \cdots, Y_N$ are independent and $\mathbb{P}[\omega_n = 1] = 1/N$. We have
$$\mathbb{P}[\omega_n = 1 | Y_n] = \begin{cases} 0, & \text{if } Y_n \geqslant 2; \\ n/N, & \text{if } Y_n = 1. \end{cases}$$
Thus
$$\mathbb{P}[\omega_{T_r} = 1] = \sum_{n=r}^{N} \mathbb{P}[\omega_n = 1, T_r = n]$$
$$= \sum_{n=r}^{N} \sum_{j_r, \cdots, j_{n-1} \neq 1} \mathbb{P}[\omega_n = 1, Y_r = j_r, \cdots, Y_{n-1} = j_{n-1}, Y_n = 1]$$
$$= \sum_{n=r}^{N} \sum_{j_r, \cdots, j_{n-1} \neq 1} \mathbb{P}[Y_r = j_r, \cdots, Y_{n-1} = j_{n-1}, Y_n = 1] \cdot$$
$$\cdot P[\omega_n = 1 | Y_r = j_r, \cdots, Y_{n-1} = j_{n-1}, Y_n = 1]$$
$$= \sum_{n=r}^{N} \frac{r-1}{r} \cdot \frac{r}{r+1} \cdot \ldots \cdot \frac{n-2}{n-1} \cdot \frac{1}{n} \cdot \frac{n}{N}$$
$$= \frac{r-1}{N} \sum_{n=r}^{N} \frac{1}{n-1} =: \varphi(r).$$
Since
$$\varphi(r) - \varphi(r+1) = \frac{1}{N}\left(1 - \sum_{n=r}^{N-1} \frac{1}{n}\right),$$
$\varphi(r)$ attains its greatest value at r^*:
$$r^* = r^*(N) = \inf\{r : \varphi(r) - \varphi(r+1) \geqslant 0\}$$
$$= \inf\left\{r : \sum_{n=r}^{N-1} \frac{1}{n} \leqslant 1\right\}.$$
Replacing summation by integral, we have $\lim_{N\to\infty} N^{-1} r^*(N) = e^{-1}$.

Theorem 4.19. When N is big enough, the optimal solution to the secretary problem is $r^*(N)/N \approx e^{-1} \approx 0.37$.

4.4 Supplements and Exercises

(1) Assume $\Omega = \{1, 2, 3\}$, $\mathscr{F}$ is the class of all subset of Ω, and on $(\Omega, \mathscr{F})$ there is a probability measure $\mathbb{P}\{i\} = p_i, \sum_i p_i = 1$. Let $\mathscr{G} = \sigma\{\{1,2\},\{3\}\}$, $X(i) = i$. Then X is a random variable from $(\Omega, \mathscr{F})$ to $(\mathbb{R}, \mathscr{B})$, but it is not measurable with respect to $\mathscr{G}$. Write out $\mathbb{E}[X|\mathscr{G}]$.

(2) Prove the assertions (1)–(5) in Theorem 4.10.

(3) If T is a stopping time, then $[T < t] \in \mathscr{F}_t$, $[T = t] \in \mathscr{F}_t$.

(4) (a) If T is a stopping time, then $\mathscr{F}_T$ is a σ-algebra.
   (b) When $T \equiv t$, $\mathscr{F}_T = \mathscr{F}_t$.
   (c) If stopping times $S \leqslant T$, then $\mathscr{F}_S \subset \mathscr{F}_T$.

(5) Assume T_1, T_2 are stopping times, then $\{T_1 < T_2\}, \{T_1 > T_2\}$, $\{T_1 \leqslant T_2\}$, $\{T_1 \geqslant T_2\}$, and $\{T_1 = T_2\}$ belong to $\mathscr{F}_{T_1} \cap \mathscr{F}_{T_2}$, and $\mathscr{F}_{T_1} \cap \mathscr{F}_{T_2} = \mathscr{F}_{T_1 \wedge T_2}$.

(6) Prove $I_{[T \geqslant S]}\mathbb{E}[X_T|X_S] = I_{[T \geqslant S]}\mathbb{E}[X_T|\mathscr{F}_{S \wedge T}], a.s.$

(7) The discrete-time Markov chain is strong Markovian.

(8) Assume $(X_n)_{n \geqslant 0}$ is a recurrent Markov chain on a countable state space $(E, \mathscr{E})$, having transition probability function $p(x, A)$. Fix a measurable set F in $\mathscr{E}$, let
$$T_1 = \inf\{n \geqslant 0 : X_n \in F\}, \quad T_{k+1} = \inf\{n > T_k : X_n \in F\}.$$
Prove $X_{T_1}, X_{T_2}, \cdots$ is a Markov chain with state space F, and write out its transition function.

(9) (Wald's identity) Assume $X_1, X_2, \cdots$ are independent, identically distributed random variables, such that $E|X_1| < \infty$. Let N be a stopping time for $\mathscr{F}_n = \sigma\{X_m, m \leqslant n\}$. Then
$$E(X_1 + \cdots + X_N) = E(N)E(X_1).$$

(10) Each Q-process is a strong Markov process.

(11) Consider a Q-process $(X_t)_{t \geqslant 0}$. Let $\tau_j = \inf\{t \geqslant 0 : X_t = j\}$ and
$$F_{ij}(t) = \mathbb{P}_i[\tau_j \leqslant t].$$
Then
$$P_{ij}(t) = \int_0^t P_{jj}(t - s)\mathrm{d}F_{ij}(s) = \int_0^t P_{ij}(t - s)\mathrm{d}F_{jj}(s).$$

(12) If $f(t, x)$ is a measurable function in (t, x), then $f(t, X(t, \omega))$ is progressively measurable.

(13) Prove the Gaussian kernel

$$\varphi(t, x, y) = \frac{1}{\sqrt{2\pi t}} \exp\left[-\frac{(y - x)^2}{2t} \right]$$

satisfies the heat equation $\partial\varphi/\partial t = \frac{1}{2}\partial^2\varphi/\partial x^2$.

PART 2
Stochastic Analysis

Chapter 5

Martingale

5.1 Definitions and Basic Properties

Denote $\mathbb{T} = \{0, 1, 2, \cdots\} = \mathbb{Z}_+$ or $\mathbb{T} = [0, \infty)$, and $\bar{\mathbb{T}} = \mathbb{T} \cup \{+\infty\}$. Assume $(\Omega, \mathscr{F}, \mathbb{P})$ is a probability space and $(\mathscr{F}_t)_{t \geq 0}$ is a filtration of increasing σ-fields: $\mathscr{F}_s \subset \mathscr{F}_t \subset \mathscr{F}$ for $s \leq t$. Let

$$\mathscr{F}_\infty = \bigvee_t \mathscr{F}_t := \sigma\Big(\bigcup_t \mathscr{F}_t\Big).$$

Definition 5.1. A family of random variables $(X_t)_{t \in \mathbb{T}}$ is **adapted**, if $X_t \in \mathscr{F}_t$ for each $t \in \mathbb{T}$. Furthermore, if X_t is integrable for each $t \in \mathbb{T}$, and

$$\mathbb{E}[X_t | \mathscr{F}_s] = (\leq, \geq) X_s, \quad s \leq t, \qquad \text{a.s.} \tag{5.1}$$

then $(X_t)_{t \in \mathbb{T}}$ is called a **(super, sub) martingale**, respectively. $(X_t)_{t \in \bar{\mathbb{T}}}$ is called a **closed (super, sub) martingale**, respectively, if in addition

$$\mathbb{E}[X_\infty | \mathscr{F}_t] = (\leq, \geq) X_t, \qquad \text{a.s.} \tag{5.2}$$

for each $t \in \mathbb{T}$. In that case, X_∞ is called the **right-closed element**.

Briefly speaking, the conditional expectations of submartingale are increasing.

If (X_t) is a submartingale, then $(-X_t)$ is a supermartingale. Therefore, the property of supermartingale can be mostly derived from that of the submartingale in this way. Thus, we often consider only the submartingale hereafter.

Lemma 5.2 (Basic properties).
(1) Martingale is closed under linear operation: $\alpha_1 X_t^{(1)} + \alpha_2 X_t^{(2)}$ for $\alpha_1, \alpha_2 \in \mathbb{R}$.
(2) Super(sub)martingale is closed under cone operation: $\alpha_1 X_t^{(1)} + \alpha_2 X_t^{(2)}$, $\alpha_1, \alpha_2 \geq 0$.

(3) The maximal of two submartingales is a submartingale.

(4) The convex function of a martingale is a submartingale. Namely, assume (X_t) is a martingale, and f is a convex function on $\mathbb{R}$. If $f(X_t)$ is integrable for each $t \in \mathbb{T}$, then $(f(X_t))_{t \in \mathbb{T}}$ is a submartingale. For example, $f(x) = |x|^\alpha$ for some $\alpha \geqslant 1$.

(5) The increasing convex function of a submartingale is a submartingale. Namely, assume (X_t) is a submartingale, and f is an increasing convex function on $\mathbb{R}$. If $f(X_t)$ is integrable for each $t \in \mathbb{T}$, then $(f(X_t))_{t \in \mathbb{T}}$ is a submartingale. For example, $f(x) = x^\alpha$, $\alpha \geqslant 1, x \geqslant 0$ or $f(x) = x^+ = x \vee 0$.

Proof We prove only (4) and (5). For (4), by convexity, we have
$$f\left(\frac{x_1 + x_2}{2}\right) \leqslant \frac{1}{2}\big(f(x_1) + f(x_2)\big).$$
In general,
$$f\left(\sum_i \lambda_i x_i\right) \leqslant \sum_i \lambda_i f(x_i), \quad \lambda_i \geqslant 0, \quad \sum_i \lambda_i = 1.$$
Then
$$f(X_s) = f\big(\mathbb{E}[X_t|\mathscr{F}_s]\big) \quad \text{(by martingale property)}$$
$$\leqslant \mathbb{E}\big[f(X_t)|\mathscr{F}_s\big] \quad \text{(by convexity and Jensen's inequality)}.$$

To prove (5), we need only to replace the first "=" by "$\leqslant$" in the previous proof.   □

Alternative proof of (4). Use Jensen's inequality for the conditional expectation. Notice that for every convex function f,
$$f(x) - f(y) \geqslant f'_+(y)(x - y).$$

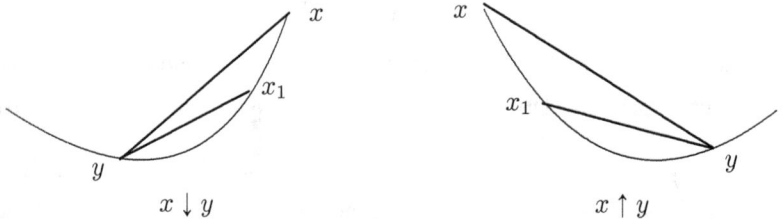

$$x \downarrow y \qquad\qquad\qquad x \uparrow y$$

Here f'_+ is the right derivative of f. Thus
$$f(X) - f(\mathbb{E}[X|\mathscr{G}]) \geqslant f'_+(\mathbb{E}[X|\mathscr{G}])(X - \mathbb{E}[X|\mathscr{G}])$$
$$\implies \mathbb{E}[f(X)|\mathscr{G}] - f(\mathbb{E}[X|\mathscr{G}]) \geqslant 0.$$
However there is the integrability problem. For this, first replace X by $XI_{[|\mathbb{E}[X|\mathscr{G}]| \leqslant N]}$, and then let $N \uparrow \infty$. And finally we take the conditional expectation.   □

5.2   Doob's Stopping Theorem

In Chapter 4, we have replaced Markov property by strong Markov property, we also want replace martingale property by strong martingale property. We begin with a simple case.

Lemma 5.3. Assume $(X_n)_{n \geqslant 0}$ is a martingale (submartingale). Let S, T be two bounded stopping times such that $S \leqslant T$. Then X_S and X_T are integrable, and

$$\mathbb{E}[X_T | \mathscr{F}_S] = (\geqslant) X_S.$$

Proof Let $N = \sup_{\omega} T(\omega) < \infty$. Then from $\mathbb{E}|X_T| \leqslant \sum_{n=1}^{N} \mathbb{E}|X_n|$, we know X_T is integrable. In the following, we prove the theorem only in the case of submartingales.

Assume first that $0 \leqslant T - S \leqslant 1$. If $A \in \mathscr{F}_S$, then

$$A[S = j][T = j + 1] = A[S = j][T > j] \in \mathscr{F}_j.$$

Thus by submartingale property,

$$\int_A (X_T - X_S) = \sum_{j=0}^{N} \int_{A[S=j][T>j]} (X_{j+1} - X_j) \geqslant 0.$$

In general, let $R_j = T \wedge (S + j)$ for $j = 0, 1, 2, \cdots, N$. Then R_j are all stopping times, such that $0 \leqslant R_{j+1} - R_j \leqslant 1$ and $R_0 = S, R_N = T$. Since

$$A \in \mathscr{F}_S \Longrightarrow A \in \mathscr{F}_{R_j}, \qquad j = 1, \cdots, N,$$

it follows from the previous proof that

$$\int_A X_S = \int_A X_{R_0} \leqslant \int_A X_{R_1} \leqslant \cdots \leqslant \int_A X_{R_N} = \int_A X_T. \quad \square$$

Theorem 5.4 (Doob's stopping theorem). Let $\bar{\mathbb{N}} = \{0, 1, 2, \cdots\} \cup \{+\infty\}$. Assume $(X_n)_{n \in \bar{\mathbb{N}}}$ is a closed (sub)martingale. Let S, T be two stopping times such that $S \leqslant T$. Then X_S and X_T are integrable, and

$$\mathbb{E}[X_T | \mathscr{F}_S] = (\geqslant) X_S.$$

Proof a) We begin with the case of martingale. For every stopping time R,

$$\mathbb{E}|X_R| = \sum_{0 \leqslant n \leqslant \infty} \int_{[R=n]} |X_n| \leqslant \sum_{0 \leqslant n \leqslant \infty} \int_{[R=n]} |X_\infty| = \mathbb{E}|X_\infty| < \infty.$$

Thus X_T and X_S are integrable. In the last inequality, we have used the fact that the absolute value of a martingale is submartingale. Note that this is not valid in the case of the submartingale.

For any $A \in \mathscr{F}_R$, we have

$$\int_A X_R = \sum_{0 \leqslant n \leqslant \infty} \int_{A[R=n]} X_n = \sum_{0 \leqslant n \leqslant \infty} \int_{A \cap [R=n]} X_\infty = \int_A X_\infty.$$

So $X_R = \mathbb{E}[X_\infty | \mathscr{F}_R]$. Further notice $\mathscr{F}_S \subset \mathscr{F}_T$. Then

$$X_S = \mathbb{E}[X_\infty | \mathscr{F}_S] = \mathbb{E}[\mathbb{E}[X_\infty | \mathscr{F}_T] | \mathscr{F}_S] = \mathbb{E}[X_T | \mathscr{F}_S].$$

b) Next, we prove the theorem for submartingale. Let

$$Y_n = \mathbb{E}[X_\infty | \mathscr{F}_n] - X_n \geqslant 0.$$

Obviously, $(Z_n = \mathbb{E}[X_\infty | \mathscr{F}_n])_{n \in \bar{\mathbb{N}}}$ is a martingale (this is the simplest martingale generated by an integrable random variable). It follows from part a) that Doob's stooping time theorem holds for the martingale (Z_n). Therefore, the stopping theorem for $(X_n)_{n \in \bar{\mathbb{N}}}$ is equivalent to that for $(Y_n)_{n \in \bar{\mathbb{N}}}$ where $\mathscr{F}_\infty = \bigvee_n \mathscr{F}_n$. But $(Y_n)_{n \in \bar{\mathbb{N}}}$ is a nonnegative supermartingale and

$$Y_\infty = \mathbb{E}[X_\infty | \mathscr{F}_\infty] - X_\infty = 0.$$

Thus, the proof of Doob's stopping theorem is reduced to consider only this nonnegative supermartingale.

First of all, for every stopping time R, we have

$$Y_R = Y_R I_{[R<\infty]} + Y_\infty I_{[R=\infty]} = Y_R I_{[R<\infty]}, \tag{5.3}$$

where we have used that $Y_\infty = 0$. Therefore

$$\mathbb{E}Y_R = \mathbb{E}\left(\lim_{n \to \infty} Y_{R \wedge n} I_{[R \leqslant n]} \right) \leqslant \mathbb{E}\left(\lim_{n \to \infty} Y_{R \wedge n} \right)$$

$$\leqslant \varliminf_{n \to \infty} \mathbb{E}Y_{R \wedge n} \quad \text{(by nonnegativeness and Fatou's lemma)}$$

$$\leqslant \mathbb{E}Y_0 < \infty. \quad \text{(by applying Lemma 5.3 to } (Y_n))$$

This shows that $0 \leqslant \mathbb{E}Y_R < \infty$, so that $X_R = \mathbb{E}[X_\infty | \mathscr{F}_R] - Y_R$ is integrable. Secondly, let $T_n = T \wedge n$, $S_n = S \wedge n$, and let $A \in \mathscr{F}_S$. Then from $A[S \leqslant n] \in \mathscr{F}_{S_n}$ and Lemma 5.3, it follows that

$$\int_{A[S \leqslant n]} Y_{S_n} \geqslant \int_{A[S \leqslant n]} Y_{T_n} \geqslant \int_{A[T \leqslant n]} Y_{T_n}.$$

Therefore

$$\int_{A[T \leqslant n]} Y_T \leqslant \int_{A[S \leqslant n]} Y_S.$$

By letting $n \uparrow \infty$, it follows from (5.3) that $\int_A Y_T \leqslant \int_A Y_S$. $\quad\square$

Corollary 5.5. Assume $(X_n)_{n \in \bar{\mathbb{Z}}_+}$ is a (sub)martingale. Let S and T be two stopping times. Then $\mathbb{E}[X_T | \mathscr{F}_S] = (\geqslant) X_{S \wedge T}$.

Proof From Theorem 5.4, we know $\mathbb{E}[X_{T \vee S} | \mathscr{F}_S] = X_S$. Thus

$$\mathbb{E}[X_T | \mathscr{F}_S] = \mathbb{E}[X_T I_{[T \leqslant S]} + X_{T \vee S} I_{[T > S]} | \mathscr{F}_S]$$
$$= X_T I_{[T \leqslant S]} + X_S I_{[T > S]}$$
$$= X_{S \wedge T}. \quad \square$$

Corollary 5.6. Assume $(X_n)_{n \geqslant 0}$ is a (sub)martingale. Then $(X_{n \wedge T})_{n \geqslant 0}$ is a (sub)martingale with respect to $(\mathscr{F}_n)_{n \geqslant 0}$.

Proof Assume $n \leqslant m$. Since $(X_k)_{k \leqslant m}$ has the right-closed element, it follows from Corollary 5.5 that

$$\mathbb{E}[X_{m \wedge T} | \mathscr{F}_n] = X_{n \wedge (m \wedge T)} = X_{n \wedge T}, \quad n \leqslant m. \quad \square$$

5.3 Fundamental Inequalities

We begin with the estimates for the distribution of the extreme values of submartingales.

Theorem 5.7. Assume $(X_n)_{n \in \mathbb{Z}_+}$ is a submartingale and $\lambda > 0$. Then

(1)

$$\lambda \mathbb{P}\left[\max_{0 \leqslant k \leqslant n} X_k \geqslant \lambda\right] \leqslant \int_{\left[\max\limits_{0 \leqslant k \leqslant n} X_k \geqslant \lambda\right]} X_n \leqslant \mathbb{E} X_n^+ \leqslant \mathbb{E}|X_n|;$$

(2)

$$\lambda \mathbb{P}\left[\min_{0 \leqslant k \leqslant n} X_k \leqslant -\lambda\right] \leqslant -\mathbb{E} X_0 + \int_{\left[\min\limits_{0 \leqslant k \leqslant n} X_k > -\lambda\right]} X_n$$
$$\leqslant \mathbb{E} X_n^+ - \mathbb{E} X_0 \leqslant \mathbb{E}|X_0| + \mathbb{E}|X_n|;$$

(3)

$$\lambda \mathbb{P}\left[\max_{1 \leqslant k \leqslant n} |X_k| \geqslant \lambda\right] \leqslant 2\mathbb{E} X_n^+ - \mathbb{E} X_0 \leqslant 2\mathbb{E}|X_n| + \mathbb{E}|X_0|.$$

Proof Denote $X_n^* = \max\limits_{0 \leqslant k \leqslant n} X_k$, $\Lambda = [X_n^* \geqslant \lambda]$. Let

$$T = \begin{cases} \min\{k : k \leqslant n, \ X_k \geqslant \lambda\}; \\ n, \quad \text{if the above set} = \varnothing. \end{cases}$$

Then T is a stopping time and $T \leqslant n$. Noting that

$$\Lambda[T = k] = [X_j < \lambda, \ 0 \leqslant j < k, \ X_k \geqslant \lambda] \in \mathscr{F}_k$$
$$\implies \Lambda \in \mathscr{F}_T.$$

For stopping times T and n, we apply Doob's stopping theorem to derive

$$\int_\Lambda X_n^+ \geq \int_\Lambda X_n \geq \int_\Lambda X_T \geq \lambda \mathbb{P}(\Lambda). \quad \text{(as on } \Lambda \text{ it holds that } X_T \geq \lambda)$$

This is just part (1).

We prove (2) in the similar way. Now, let

$$\Lambda = \left[\min_{0 \leq k \leq n} X_n \leq -\lambda \right]$$

and

$$T = \begin{cases} \min\{k : k \leq n,\ X_k \leq -\lambda\}; \\ n, \quad \text{if the above set } = \varnothing. \end{cases}$$

Then

$$\mathbb{E}X_0 \leq \mathbb{E}X_T = \int_\Lambda X_T + \int_{\Lambda^c} X_T \leq -\lambda \mathbb{P}(\Lambda) + \int_{\left[\min_{0 \leq k \leq n} X_k > -\lambda \right]} X_n.$$

Part (3) of the theorem follows immediately from parts (1)-(2). $\square$

Remark 5.8. In the infinite case, the corresponding assertions in Theorem 5.7 hold. For instant, part (1) of the theorem becomes

$$\lambda \mathbb{P}\left[\max_{1 \leq k < \infty} X_k \geq \lambda \right] \leq \sup_n \mathbb{E}X_n^+.$$

Indeed, from part (1) of the theorem, we have

$$\lambda \mathbb{P}\left[\max_{1 \leq k \leq n} X_k \geq \lambda \right] \leq \mathbb{E}X_n^+.$$

Since

$$X_n^* = \max_{1 \leq k \leq n} X_k \uparrow \quad \text{and} \quad I_{[\lambda,\infty)}(f_n) \uparrow I_{[\lambda,\infty)}(f_\infty)$$

as $n \uparrow$, it follows from the monotone convergence theorem that

$$\lambda \mathbb{P}\left[\max_{1 \leq k < \infty} X_k \geq \lambda \right] \leq \sup_n \mathbb{E}X_n^+.$$

Corollary 5.9 (Kolmogorov's inequality). Assume $(X_n)_{n \geq 0}$ is a martingale, and $X_n \in L^p\ (p \geq 1)$ for every n, i.e.

$$\int |X_n|^p d\mathbb{P} < \infty.$$

Then

$$\mathbb{P}\left[\max_{0 \leq k \leq n} |X_k| \geq \lambda \right] \leq \mathbb{E}|X_n|^p / \lambda^p$$

for every n.

Proof If (X_n) is a martingale, then $(|X_n|^p)$ is a submartingale, and we apply Theorem 5.7 to derive the assertion.   □

Now, assume $(X_n)_{n \geqslant 1}$ are independent and identically distributed (i.i.d.), and $\mathbb{E}X_n = 0$ and $\mathbb{E}X_n^2 < \infty$ for every n. Let $\mathscr{F}_n = \sigma(X_k : k \leqslant n)$. Then

$$\begin{aligned}
\mathbb{E}[S_{n+1}|\mathscr{F}_n] &= \mathbb{E}[X_1 + \cdots + X_{n+1}|\mathscr{F}_n] \\
&= S_n + \mathbb{E}[X_{n+1}|\mathscr{F}_n] \\
&= S_n + \mathbb{E}[X_{n+1}] \quad \text{(by independence)} \\
&= S_n \quad \text{(zero mean).}
\end{aligned}$$

Thus $(S_n, \mathscr{F}_n)$ is a martingale, and $(|S_n|^2)$ is a submartingale. Therefore

$$\mathbb{P}\left[\max_{1 \leqslant k \leqslant n} |S_k| \geqslant \lambda \right] \leqslant \lambda^{-2}\mathbb{E}S_n^2.$$

This is the classical Kolmogorov's inequality (1929). K. L. Chung have claimed in his textbook ([20]) that this is an excellent inequality, whose original proof was an example using the exquisite skills in probability theory.

The importance of martingale (supermartingale, submartingale) is not completely due to the fact that it consists a class of stochastic processes motivated from practice. In other words, it is not due to its direct applications in practice, but more to its concise property which makes an important tool in theoretical research. On the other hand, as we have seen from the above example, the key for application is how to find a martingale from the original stochastic process (e.g. i.i.d. (X_n)). We will come back to this topic later.

The following result describes the moments for the extreme values of a nonnegative submartingale. For this, let

$$\| X \|_p := \| X \|_{L^p} := \left(\int |X|^p \mathrm{d}\mathbb{P} \right)^{1/p}$$

for a random variable and $p \geqslant 1$.

Theorem 5.10 (Doob's inequality). Assume $(X_n)_{n \geqslant 0}$ is a nonnegative submartingale, and let $X_n^* = \max_{0 \leqslant k \leqslant n} X_k$. Assume further φ is a right-continuous function on $\mathbb{R}_+$ with $\varphi(0) = 0$. Then

$$\mathbb{E}\varphi(X_n^*) \leqslant \mathbb{E}\left[X_n \int_0^{X_n^*} \frac{1}{\lambda} \mathrm{d}\varphi(\lambda) \right].$$

In particular,

$$\|X_n^*\|_p \leqslant \begin{cases} q\| \, X_n \, \|_p, & \text{if } p > 1; \\ \frac{e}{e-1}[1 + \mathbb{E}(X_n \log^+ X_n)], & \text{if } p = 1, \end{cases}$$

where q is the conjugate number of p: $1/p + 1/q = 1$, and $\log^+ x = \log(e \vee x)$.

Proof From Fubini's theorem and (1) of Theorem 5.7, it follows that

$$\mathbb{E}\varphi(X_n^*) = \int_\Omega \int_{[0,X_n^*]} d\varphi(\lambda) d\mathbb{P} = \int_0^\infty \mathbb{P}[X_n^* \geqslant \lambda] d\varphi(\lambda)$$

$$\leqslant \int_0^\infty \frac{1}{\lambda} d\varphi(\lambda) \int_{[X_n^* \geqslant \lambda]} X_n d\mathbb{P}$$

$$= \mathbb{E}\left[X_n \int_0^{X_n^*} \frac{1}{\lambda} d\varphi(\lambda) \right] \quad \text{(Fubini's theorem)}.$$

Now take $\varphi(\lambda) = (\lambda - 1)^+$. As

$$\int_0^X \frac{1}{\lambda} d(\lambda - 1)^+ = \int_1^{X \vee 1} \frac{1}{\lambda} d\lambda = \log(X \vee 1) = \log^+ X,$$

we have

$$\mathbb{E}(X_n^* - 1) \leqslant \mathbb{E}[(X_n^* - 1)^+] \leqslant \mathbb{E}[X_n \log^+ X_n^*].$$

By using the following element inequality,

$$\log x \leqslant x/e \quad (x \geqslant 0) \implies a \log^+ b \leqslant a \log^+ a + b/e,$$

(When $b \leqslant 1$, this is trivial. When $b > 1$, by letting $x = b/a$, the latter inequality comes from the former one.) we have

$$\mathbb{E}[X_n \log^+ X_n^*] \leqslant \mathbb{E}[X_n \log^+ X_n + X_n^*/e].$$

From this and the previous inequality, we obtain the last assertion. Finally, take $\varphi(\lambda) = \lambda^p$ $(p > 1)$. Then it follows from Hölder's inequality that

$$\mathbb{E}[X_n^{*p}] \leqslant \frac{p}{p-1} \mathbb{E}[X_n X_n^{*(p-1)}] \leqslant q(\mathbb{E}X_n^p)^{1/p}(\mathbb{E}X_n^{*p})^{1/q}.$$

Without loss of generality, assume $\|X_n\|_p < \infty$. As $(X_n^*)_{n \geqslant 0}$ is still a submartingale, we have

$$\mathbb{E}X_n^{*p} \leqslant \sum_{k=1}^n \mathbb{E}X_k^p \leqslant n\mathbb{E}X_n^p < \infty.$$

Thus, if $\mathbb{E}X_n^{*p} \neq 0$, then divided both sides by $\left[\mathbb{E}(X_n^{*p})\right]^{1/q}$, we get the assertion for $p > 1$. If $\mathbb{E}X_n^{*p} = 0$, then $\mathbb{E}X_n^p = 0$, so the inequality becomes equality.   □

5.4 Convergence Theorems

In the following, we are going to study the convergence theorems for submartingales. For this, we need some preparation.

Recall in the non-random case, if a real number sequence $\{x_n\}$ diverges, then

$$\varliminf_n x_n < \varlimsup x_n.$$

So there exist $a, b \in \mathbb{R}$ such that

$$\varliminf_n x_n < a < b < \varlimsup_n x_n.$$

From this, we see that $\{x_n\}$ will jump infinitely times from the places lower than a to the places upper than b. In details, set

$$s_1 = \inf\{n \geqslant 0 : x_n < a\}$$
$$t_1 = \inf\{n > s_1 : x_n > b\}$$
$$s_2 = \inf\{n > t_1 : x_n < a\}$$
$$\vdots$$

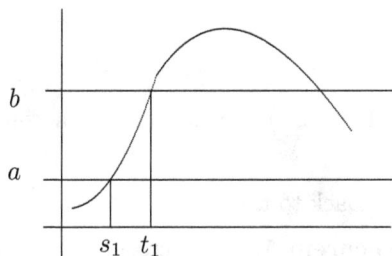

For example, if $t_1 < \infty$, then from s_1 to t_1, $\{x_n\}_{n\geqslant 0}$ upcrosses the interval $[a, b]$ once. Thus, sequence $\{x_n\}_{n\geqslant 0}$ diverges iff $\{x_n\}_{n\geqslant 0}$ upcrosses some interval $[a, b]$ infinite times.

Now we consider submartingale $(X_n)_{0\leqslant n\leqslant N}$ with fixed N. We study the times that (X_n) "upcrosses". Since $\{(X_n - a)^+\}_{0\leqslant n\leqslant N}$ is still a submartingale, we may and do assume that $(X_n)_{0\leqslant n\leqslant N}$ is a nonnegative submartingale, and consider only in the case of $a = 0$. Let

$$S_1 = \min\{n : N \geqslant n \geqslant 0, X_n = 0\},$$
$$T_1 = \min\{n : N \geqslant n \geqslant S_1, X_n \geqslant b\},$$

$$\vdots$$

$$S_k = \min\{n : N \geqslant n \geqslant T_{k-1}, X_n = 0\},$$
$$T_k = \min\{n : N \geqslant n \geqslant S_k, X_n \geqslant b\},$$

$$\vdots$$

As usual, define in addition $\min \varnothing = N$. Clearly, if the upcrossing times $V_0^b(X, N) =: j > 0$, then

$$(X_{T_1} - X_{S_1}) + \cdots + (X_{T_j} - X_{S_j}) \geqslant jb. \tag{5.4}$$

This also holds when $j = 0$, since the summation on empty set is 0 by convention. Insert $j = V_0^b(X, N)$ into the previous inequality and take the expectation to derive

$$\mathbb{E}(X_{T_1} - X_{S_1}) + \cdots + \mathbb{E}\left(X_{T_{V_0^b(X,N)}} - X_{S_{V_0^b(X,N)}}\right) \geqslant b\mathbb{E}V_0^b(X, N).$$

This is too complicated to use directly. However, by submartingale property, we have

$$\mathbb{E}(X_{S_1} - X_0) \geqslant 0,$$
$$\mathbb{E}(X_{S_2} - X_{T_1}) \geqslant 0,$$
$$\vdots$$
$$\mathbb{E}\left(X_{S_{V_0^b(X,N)}} - X_{T_{V_0^b(X,N)-1}}\right) \geqslant 0,$$
$$\mathbb{E}\left(X_N - X_{T_{V_0^b(X,N)}}\right) \geqslant 0.$$

Taking these inequalities into account, from (5.4) we obtain that

$$\mathbb{E}X_N - \mathbb{E}X_0 \geqslant b\,\mathbb{E}V_0^b(X, N).$$

Back to the original case, we have proved the following results.

Theorem 5.11 (Upcrossing inequality). Assume that $X = (X_n)_{0 \leqslant n \leqslant N}$ is a submartingale. Let $V_a^b(X, N)$ be the times of X upcrossing $[a, b]$, that is the times of $(X - a)^+$ upcrossing $[0, b - a]$. Then

$$\mathbb{E}V_a^b(X, N) \leqslant \frac{1}{b - a}\left[\mathbb{E}(X_N - a)^+ - \mathbb{E}(X_0 - a)^+\right] \leqslant \frac{1}{b - a}[\mathbb{E}X_N^+ + |a|].$$

For submartingale $(X_n)_{n \in \mathbb{Z}_+}$, from the "upcrossing" inequality, we know that $V_a^b(X, N) \uparrow$ some V_a^b when $N \to \infty$. On the other hand, since (X_n^+) is also a submartingale, we have $\mathbb{E}X_N^+ \uparrow \sup_n \mathbb{E}X_n^+$. Thus, if $\sup_n \mathbb{E}X_n^+ < \infty$, then

$$\mathbb{E}V_a^b \leqslant \left(\sup_n X_n^+ + |a|\right)/(b - a) < \infty$$
$$\Longrightarrow V_a^b < \infty, \quad \text{a.s.}$$

Denote the exception set by Λ_a^b and let $\Lambda = \bigcup_{a < b,\, a, b \in \mathbb{Q}} \Lambda_a^b$. Then $\mathbb{P}(\Lambda) = 0$, and on Λ^c,

$$\varliminf_n X_n = \varlimsup_n X_n =: X_\infty.$$

This implies the following fundamental theorem.

Theorem 5.12. Assume $(X_n)_{n \geqslant 0}$ is a submartingale such that $\sup_n \mathbb{E} X_n^+ <$
∞ $\left(\Longleftrightarrow \sup_n \mathbb{E}|X_n| < \infty \right)$. Then there exists $X_\infty \in \mathscr{F}_\infty = \bigvee_n \mathscr{F}_n$ such that
$X_n \xrightarrow{\text{a.s.}} X_\infty$ and $\mathbb{E}|X_\infty| < \infty$.

Proof Since (X_n) is a submartingale,

$$\mathbb{E}|X_n| = 2\mathbb{E} X_n^+ - \mathbb{E} X_n \leqslant 2\mathbb{E} X_n^+ - \mathbb{E} X_1.$$

This shows "$\Longleftrightarrow$". Moreover, by Fatou's lemma,

$$\infty > \sup_n \mathbb{E}|X_n| = \varliminf_{n \to \infty} \mathbb{E}|X_n| \geqslant \mathbb{E}|X_\infty|. \quad \square$$

Remark 5.13. Since (X_n) is a submartingale, we have $\mathbb{E} X_N \uparrow \sup_n \mathbb{E} X_n \leqslant$
$\sup_n \mathbb{E} X_n^+$ when $N \to \infty$. From this we see that the assumptions in Theorem
5.12 is a little stronger than the natural condition: $\sup_n \mathbb{E} X_n < \infty$.

When studying Doob's stopping theorem, we have ever assumed that
$(X_n)_{n \in \bar{\mathbb{Z}}_+}$ is a martingale (submartingale) with the right-closed element X_∞.
Note that for $(X_n)_{n \in \mathbb{Z}_+}$ as in Theorem 5.12, we have proved $X_n \xrightarrow{\text{a.s.}} X_\infty$.
How can we make $(X_n)_{n \in \bar{\mathbb{Z}}_+}$ be a closed submartingale? What we need to
prove is

$$\mathbb{E}[X_\infty | \mathscr{F}_n] \geqslant X_n, \quad n \geqslant 0.$$

This is equivalent to

$$\int_A X_\infty \geqslant \int_A X_n, \quad A \in \mathscr{F}_n, \, n \geqslant 0.$$

However we already have

$$\int_A X_m \geqslant \int_A X_n, \quad A \in \mathscr{F}_n, \, m \geqslant n \geqslant 0.$$

Thus, to keep the right-closeness, we need

$$\int_A X_\infty \geqslant \varliminf_m \int_A X_m \geqslant \int_A X_n, \quad A \in \mathscr{F}_n, n \geqslant 0.$$

If $\{X_n\}$ are nonnegative, then Fatou's lemma implies the reverse inequality.
Therefore we need more conditions. Condition "$|X_n| \leqslant Y$ integrable" is
sufficient, but as we will see, this is stronger than the following condition:

$$\int_A X_m \xrightarrow[m \to \infty]{} \int_A X_\infty, \quad A \in \mathscr{F}_n, \, n \geqslant 0.$$

This indeed requires that the above equation holds for every $A \in \bigcup_n \mathscr{F}_n$. A slight strengthening of this condition becomes

$$\int_A X_m \to \int_A X_\infty, \qquad \forall A \in \mathscr{F}$$

$$\iff (X_n)_{n \geqslant 0} \text{ is a } \textbf{uniformly integrable} \text{ family.}$$

See [49; p. 118]. Recall that $X \in L^1(\mathbb{P})$ is equivalent to

$$\lim_{\alpha \to \infty} \int_{[|X| \geqslant \alpha]} |X| =: \mathbb{E}(|X|; |X| \geqslant \alpha) \longrightarrow 0.$$

Similarly, we introduce the following concept.

Definition 5.14. We call a family $\mathscr{H}$ of random variables **uniformly integrable**, if

$$\lim_{\alpha \to \infty} \sup_{X \in \mathscr{H}} \mathbb{E}[|X|; |X| \geqslant \alpha] = 0.$$

Obviously, if $|X_n| \leqslant Y$ and Y is integrable, then (X_n) is uniformly integrable.

Theorem 5.15 (Criterion for uniform integrability). In order that $\mathscr{H} \subset L^1$ is uniformly integrable, it is necessary and sufficient that
   (1) L^1-uniform boundedness: $\sup\limits_{X \in \mathscr{H}} \mathbb{E}|X| < \infty$; and
   (2) Equicontinuity in probability: for every $\varepsilon > 0$, there exists $\delta > 0$ such that

$$\mathbb{P}(A) < \delta \implies \sup_{X \in \mathscr{H}} \mathbb{E}(|X|; A) < \varepsilon.$$

Theorem 5.16 (Criterion for L^1-convergence). Assume that $(X_n)_{n \geqslant 0} \subset L^1$. In order that

$$X_n \xrightarrow{L^1} X_\infty \left(\iff \int |X_n - X_\infty| \to 0 \right), \qquad n \to \infty,$$

it is necessary and sufficient that $(X_n)_{n \geqslant 0}$ are uniformly integrable and $X_n \xrightarrow{\mathbb{P}} X_\infty$.

Proof For proofs of these two theorems, refer to [72; §6.3], [47; Appendix (2)] or [20; p. 101].   □

Let us come back to the proof of the right-closedness theorem. Assume that $(X_n)_{n \geqslant 0}$ are uniformly integrable. By

$$\mathbb{E}|X_n| \leqslant \alpha + \int_{[|X_n| \geqslant \alpha]} |X_n|$$

and uniform integrability, we know that $\sup_n \mathbb{E}|X_n| < \infty$. Thus it follows from Theorem 5.12 that $X_n \xrightarrow{\text{a.s.}} X_\infty \in L^1$. From this and uniform integrability, we have

$$X_n \xrightarrow[L^1]{\text{a.s.}} X_\infty \implies X_n I_A \xrightarrow[L^1]{} X_\infty I_A, \quad A \in \mathscr{F}$$

$$\implies \int_A X_n \to \int_A X_\infty, \quad A \in \mathscr{F}.$$

Sometimes, we need to consider a decreasing filtration of σ-fields $(\mathscr{F}_n)_{n \geq 0}$: $\mathscr{F}_m \subset \mathscr{F}_n$ for $m \geq n$. In this case, we set $\mathscr{G}_{-n} = \mathscr{F}_n$ for $n \geq 0$. Then $(\mathscr{G}_{-n})_{n \geq 0}$ becomes an increasing σ-field filtration: if $-m \leq -n$, then $\mathscr{G}_{-m} \subset \mathscr{G}_{-n}$.

Definition 5.17. Suppose $(X_n)_{n \geq 0}$ is an $(\mathscr{F}_n)_{n \geq 0}$-adapted and integrable stochastic process, and $(\mathscr{F}_n)$ is decreasing. Let $\mathscr{G}_{-n} = \mathscr{F}_n$ and $Y_{-n} = X_n$ for $n \geq 0$. Then $(X_n)_{n \geq 0}$ is called an $(\mathscr{F}_n)$ **reverse martingale (reverse submartingale)**, if $(Y_{-n})_{n \geq 0}$ is a $(\mathscr{G}_{-n})$ martingale (submartingale).

Let us consider the convergence of the reverse (sub)martingale. Since (Y_{-n}^+) is a $(\mathscr{G}_{-n})$ submartingale,

$$\sup_{n \geq 0} \mathbb{E} Y_{-n}^+ \leq \mathbb{E} Y_0^+ = \mathbb{E} X_0^+ < \infty.$$

Then there exists $X_{-\infty}$ such that

$$X_n = Y_{-n} \longrightarrow X_{-\infty}, \quad \text{a.s.}$$

On the other hand, from $X_n \in \mathscr{F}_n$ and $\mathscr{F}_n \downarrow$, we have

$$X_{-\infty} \in \bigwedge_{n \geq 0} \mathscr{F}_n := \bigcap_{n \geq 0} \mathscr{F}_n.$$

But since $\mathbb{E}[X_n | \mathscr{F}_m] \geq X_m$ for $m > n$, we have $\mathbb{E} X_n \geq \mathbb{E} X_m$. That is, $\mathbb{E} X_n \downarrow$ as $n \uparrow$. Thus we need a better condition "$\lim_{n \to \infty} \mathbb{E} X_n > -\infty$" to guarantee that integrability for $X_{-\infty}$.

Theorem 5.18. Suppose that $(X_n, \mathscr{F}_n)_{n \geq 0}$ is a reverse submartingale. If $\lim_{n \to \infty} \mathbb{E} X_n > -\infty$, then $(X_n)_{n \geq 0}$ is uniformly integrable and there exists $X_{-\infty} \in \bigcap_n \mathscr{F}_n$ such that

$$X_n \xrightarrow[L^1]{\text{a.s.}} X_{-\infty}.$$

Therefore $(X_n, \mathscr{F}_n)_{n \in \bar{\mathbb{Z}}_+}$ is also a reverse submartingle.

Proof We need only prove the uniform integrability for $(X_n)_{n \geq 0}$. For $\varepsilon > 0$, choose k such that

$$\mathbb{E}X_k - \lim_{n \to \infty} \mathbb{E}X_n < \varepsilon.$$

Then from $\mathbb{E}[X_k | \mathscr{F}_n] \geq X_n$ for $n \geq k$, we have

$$
\begin{aligned}
\mathbb{E}[|X_n|; |X_n| \geq \alpha] &= \mathbb{E}[X_n; X_n \geq \alpha] - \mathbb{E}[X_n; X_n \leq -\alpha] \\
&= \mathbb{E}[X_n; X_n \geq \alpha] + \mathbb{E}[X_n; X_n > -\alpha] - \mathbb{E}X_n \\
&\leq \mathbb{E}[X_k; X_n \geq \alpha] + \mathbb{E}[X_n; X_n > -\alpha] - \mathbb{E}X_k + \varepsilon \\
&\leq \mathbb{E}[X_k; X_n \geq \alpha] - \mathbb{E}[X_k; X_n \leq -\alpha] + \varepsilon \\
&\leq \mathbb{E}[|X_k|; X_n \geq \alpha] + \mathbb{E}[|X_k|; X_n \leq -\alpha] + \varepsilon \\
&= \mathbb{E}[|X_k|; |X_n| \geq \alpha] + \varepsilon.
\end{aligned}
$$

But when $\alpha \to \infty$,

$$\mathbb{P}[|X_n| \geq \alpha] \leq \frac{1}{\alpha}\mathbb{E}|X_n| = \frac{1}{\alpha}(2\mathbb{E}X_n^+ - \mathbb{E}X_n) \leq \frac{1}{\alpha}(2\mathbb{E}X_k^+ - \lim_{n \to \infty} \mathbb{E}X_n) \to 0$$

holds uniformly in $n \geq k$.

From Theorem 5.15, it follows that when α is sufficiently large,

$$\sup_{n \geq k} \mathbb{E}[|X_k|; |X_n| \geq \alpha] \leq \varepsilon.$$

Thus when α is sufficiently large,

$$\sup_n \mathbb{E}[|X_n|; |X_n| \geq \alpha] \leq \sup_{n < k} \mathbb{E}[|X_n|; |X_n| \geq \alpha] + \sup_{n \geq k} \mathbb{E}[|X_n|; |X_n| \geq \alpha]$$

$$\leq 3\varepsilon. \quad \square$$

5.5 Continuous-Time (Sub/Super)Martingale

When studying the continuous-time (sub)martingales, we naturally expect reducing them to the discrete-time ones. Recall that when studying the strong Markov property, a natural condition is that the process is right-continuous when passing discrete stopping times to continuous ones. In this case,

$$T_n \downarrow T \Longrightarrow X_{T_n} \to X_T.$$

Therefore, the right-continuity plays an important role, for example, for Doob's stopping theorem:

$$\mathbb{E}[X_t | \mathscr{F}_s] = (\geqslant)X_s \Longrightarrow \begin{cases} \mathbb{E}[X_T | \mathscr{F}_S] = (\geqslant)X_S, & \text{if } T \geqslant S; \\ \mathbb{E}[X_T | \mathscr{F}_S] \geqslant X_{S \wedge T}, & \text{in general.} \end{cases}$$

Furthermore, when studying the strong Markov property, we also need the Feller property: the mapping $x \to \mathbb{E}_x f(X_t)$ is continuous for every $f \in C_b(\mathbb{R})$. Here, we replace it by

$$T_n \downarrow T, \qquad \mathscr{F}_{T_n} \downarrow \mathscr{F}_T.$$

This means that $(\mathscr{F}_t)_{t \geqslant 0}$ is right-continuous: $\mathscr{F}_t = \bigcap_{s > t} \mathscr{F}_s$. If not, we put $\mathscr{F}_{t+} = \bigcap_{s > t} \mathscr{F}_s$ instead of $\mathscr{F}_t$, then the σ-field filter $(\mathscr{F}_{t+})_{t \geqslant 0}$ becomes right-continuous. Having these ideas, we can easily extend the results of the discrete-time (sub)martingales to that of the continuous-time ones.

As an example, we prove the following:

Theorem 5.19. Suppose that $(X_t)_{t \geqslant 0}$ is an a.s. right-continuous (sub)martingale. Then $t \to X(t, \omega)$ is bounded on every finite interval for almost surely ω.

Proof It suffice to prove that $\sup_{t \in [0,n]} |X_t|$ is bounded with probability 1 for every $n \geqslant 1$. This, by the right-continuity, is equivalent to

$$\sup_{t \in D \cap [0,n]} |X_t| = \sup_{t \in [0,n]} |X_t|$$

bounded with probability 1, where D is any countable dense subset of $[0, \infty)$. Next, set $D_n = D \cap [0, n]$. Let U_N be an arbitrary finite subset of D_n, and set

$$U_N = \{t_0 < t_1 < \cdots < t_N\}.$$

Since $(Y_k = X_{t_k})_{0 \leqslant k \leqslant N}$ is a submartingale, we have

$$\mathbb{P}\left[\max_{0 \leqslant k \leqslant N} |X_{t_k}| \geqslant \alpha \right] \leqslant \frac{2}{\alpha} \mathbb{E}|X_{t_N}| \leqslant \frac{2}{\alpha} \mathbb{E}|X_n|.$$

By letting $U_N \uparrow D_n$, we obtain

$$\mathbb{P}\left[\sup_{t \in D_n} |X_t| \geq \alpha\right] \leq \frac{2}{\alpha}\mathbb{E}|X_n|.$$

Next, by letting $\alpha \to \infty$,

$$\mathbb{P}\left[\sup_{t \in D_n} |X_t| = \infty\right] = 0. \quad \square$$

Recall that for each submartingale $(X_t)_{t \geq 0}$, it is assumed in advance to be adapted:

$$X_t \in \mathscr{F}_t, \qquad t \geq 0.$$

As mentioned before, when $(\mathscr{F}_t)$ is not right-continuous, we replace $\mathscr{F}_t$ by $\mathscr{F}_{t+}$, so that we can assume that $(\mathscr{F}_t)$ is right-continuous. In this case, the adapted condition "$X_t \in \mathscr{F}_{t+}$" is trivial. In the sequel, we always assume that $(\mathscr{F}_t)_{t \geq 0}$ is right-continuous. Next, when (X_t) is not right-continuous, what can we do?

Definition 5.20. $(\widehat{X}_t)$ is called a modification of (X_t), if $\widehat{X}_t = X_t$ a.s. for every t,

Theorem 5.21. Suppose that (X_t) is a submartingale and $(\mathscr{F}_t)$ is right-continuous. Then (X_t) has a right-continuous modification if and only if $\mathbb{E}X_t$ is right-continuous.

Proof See [47; p. 199, Theorem 3] or [57; p. 65, Theorem II2.9]. $\square$

5.6   Two Applications of Martingale Theory

Strong Law of Large Number

A typical application of the martingale theory—the classical strong law of large number—was regarded as one of the most important achievements of the martingale theory (J. Doob, 1949).

Theorem 5.22. Suppose $(X_n)_{n \geq 1}$ are independent and identically-distributed such that $\mathbb{E}|X_1| < \infty$. Then

$$\frac{1}{n}\sum_{k=1}^{n} X_k \xrightarrow{\text{a.s.}} \mathbb{E}X_1, \qquad n \to \infty.$$

Proof Without loss of generality, assume that $\mathbb{E}X_1 = 0$, so that $S_n = \sum_{k=1}^{n} X_k$ is an $\mathscr{F}_n := \sigma(S_k : k \leq n)$ martingale. Set $\mathscr{G}_n = \sigma(S_m : m \geq n)$, then $\mathscr{G}_n \downarrow \mathscr{G} := \bigcap_{n \geq 1} \mathscr{G}_n$. Fix $k : 1 \leq k \leq n$. Consider the martingale $Z_n = $

$\mathbb{E}[X_k|\mathscr{G}_n]$. Then from the convergence theorem for the reverse martingale (Theorem 5.18), it follows that

$$\mathbb{E}Z_n = \mathbb{E}X_k > -\infty, \qquad Z_n \xrightarrow[L^1]{\text{a.s.}} Z_{-\infty} \in \mathscr{G}.$$

In particular,

$$\int_A \mathbb{E}[X_k|\mathscr{G}] = \int_A X_k = \int_A Z_n \to \int_A Z_{-\infty}, \qquad A \in \mathscr{G}.$$

Therefore $Z_{-\infty} = \mathbb{E}[X_k|\mathscr{G}]$. On the other hand, since $\mathscr{G}_n = \sigma(S_n, X_m : m \geq n+1)$, and noting that $(X_m, m > n)$ is independent of (X_k, S_n), we have for all $m_1, \cdots, m_\ell \geq n+1$,

$$\mathbb{E}\{g(S_n)f_1(X_{m_1}) \cdots f_\ell(X_{m_\ell})\mathbb{E}(X_k|S_n, X_m : m \geq n+1)\}$$
$$= \mathbb{E}\{\mathbb{E}[X_k g(S_n)f_1(X_{m_1}) \cdots f_\ell(X_{m_\ell})|S_n, X_m : m \geq n+1)]\}$$
$$= \mathbb{E}[X_k g(S_n)] \, \mathbb{E}[f_1(X_{m_1}) \cdots f_\ell(X_{m_\ell})]. \quad \text{(by independence)}$$

And by independence again,

$$\mathbb{E}\{g(S_n)f_1(X_{m_1}) \cdots f_\ell(X_{m_\ell})\mathbb{E}[X_k|S_n]\}$$
$$= \mathbb{E}\{f_1(X_{m_1}) \cdots f_\ell(X_{m_\ell})\mathbb{E}[X_k g(S_n)|S_n]\}$$
$$= \mathbb{E}\{f_1(X_{m_1}) \cdots f_\ell(X_{m_\ell})\}\mathbb{E}[X_k g(S_n)].$$

Furthermore,

$$\mathbb{E}[X_k|\mathscr{G}_n] = \mathbb{E}[X_k|S_n, X_m : m \geq n+1] = \mathbb{E}[X_k|S_n] = \mathbb{E}[X_1|S_n].$$

The latter equality comes from the fact that X_1 and X_k are in the symmetric positions:

$$\mathbb{E}[f(S_n)X_1] = \mathbb{E}[f(S_n)X_k].$$

This can be proved by the monotone class theorem. Make summation from 1 to n in k in the previous equation to derive

$$S_n/n = \mathbb{E}(X_1|S_n) = \mathbb{E}(X_1|\mathscr{G}_n).$$

Using Theorem 5.18 again, we obtain

$$\mathbb{E}[X_1|\mathscr{G}_n] \xrightarrow[L^1]{\text{a.s.}} \mathbb{E}[X_1|\mathscr{G}], \qquad n \to \infty.$$

Finally, from $\lim_n S_n/n = \lim_n \sum_{j=m}^n X_j/n$, it follows that $\lim_n S_n/n$ is tail-measurable (Kolmogorov's 0-1 law). Thus it must be a constant, which must be $\mathbb{E}[\mathbb{E}[X_1|\mathscr{G}]] = \mathbb{E}X_1 = 0$. $\quad\square$

Martingale Description for Markov Chain

Let us come back to the continuous-time Markov chain (Q-process). Suppose E is countable and $(X_t)_{t \geqslant 0}$ is a Markov chain with time-homogeneous transition probability matrix $(P_{ij}(t))$:

$$\mathbb{P}[X_t = j | X_0 = i] = p_{ij}(t).$$

Assume its Q-matrix $Q = (q_{ij})$ is totally stable:

$$q_i = -q_{ii} < \infty, \quad i \in E,$$

and conservative:

$$\sum_{j \neq i} q_{ij} = q_i, \quad i \in E.$$

Assume in addition that the Q-process is unique, so that $(p_{ij}(t))$ satisfies the forward equation

$$p_{ij}'(t) = \sum_{k \in E} p_{ik}(t) q_{kj}.$$

Take integral in t on both sides to get that

$$
\begin{aligned}
p_{ij}(t) - p_{ij}(s) &= \int_s^t \sum_k p_{ik}(u) q_{kj} \mathrm{d}u \\
&= \int_s^t \sum_{k \neq j} p_{ik}(u) q_{kj} \mathrm{d}u - \int_s^t p_{ij}(u) q_j \mathrm{d}u.
\end{aligned}
\tag{5.5}
$$

Denote by $\mathbb{P}_i$ the probability measure when the process starts at i, that is,

$$\mathbb{P}_i[X_t = j] = \mathbb{P}[X_t = j | X_0 = i],$$

and by $\mathbb{E}_i$ the corresponding expectation. Then

$$\mathbb{E}_i f(X_t) = \mathbb{E}[f(X_t) | X_0 = i] = \sum_k f_k \mathbb{P}[X_t = k | X_0 = i] = \sum_k p_{ik}(t) f_k.$$

Next, let $Qf(i) = \sum_j q_{ij} f_j$. Then by taking $f = I_j$ with $I_j(k) = \delta_{jk}$ for $k \in E$, we have

$$q_{kj} = (QI_j)(k) = \sum_{\ell} q_{k\ell} I_j(\ell),$$

$$\sum_k p_{ik}(u) q_{kj} = \sum_k \mathbb{E}_i[q_{kj} : X_u = k] = \sum_k \mathbb{E}_i[QI_j(k) : X_u = k]$$

$$= \mathbb{E}_i(QI_j(X_u)).$$

From (5.5) and Fubini's theorem, it follows that

$$\mathbb{E}_i I_j(X_t) - \mathbb{E}_i I_j(X_s) = \int_s^t \mathbb{E}_i(QI_j(X_u))du = \mathbb{E}_i\left[\int_s^t QI_j(X_u))du\right],$$

or

$$\mathbb{E}_i\left[I_j(X_t) - I_j(X_s) - \int_s^t QI_j(X_u)du\right] = 0.$$

Now let $(\mathscr{F}_t)$ be the natural σ-field filtration. Then by the Markov property

$$\mathbb{E}_i\left[I_j(X_t) - \int_0^t QI_j(X_u)du\Big|\mathscr{F}_s\right]$$

$$= \mathbb{E}_i(I_j(X_t)|X_s) - \mathbb{E}_i\left[\int_s^t QI_j(X_u)du\Big|X_s\right] - \int_0^s QI_j(X_u)du$$

$$= \mathbb{E}_i[(I_j(X_s)|X_s) + \mathbb{E}_i\left[I_j(X_t) - I_j(X_s) - \int_s^t QI_j(X_u)du\Big|X_s\right]$$

$$\qquad - \int_0^s QI_j(X_u)du$$

$$= I_j(X_s) - \int_0^s QI_j(X_u)du + \mathbb{E}_{X(s)}\left[I_j(X_{t-s}) - I_j(X_0) - \int_0^{t-s} QI_j(X_u)du\right]$$

$$= I_j(X_s) - \int_0^s QI_j(X_u)du.$$

This means that

$$\left(I_j(X_t) - \int_0^t QI_j(X_u)du, \ \mathscr{F}_t\right)$$

is a $\mathbb{P}_i$-martingale. In general, if f depends only on finite states in E (i.e. with compact support), which is denoted by $f \in C_0(E)$, then

$$\left(f(X_t) - \int_0^t Qf(X_u)du, \ \mathscr{F}_t\right)$$

is a $\mathbb{P}_i$-martingale. Simply denote it by

$$X_t^f = f(X_t) - \int_0^t Qf(X_u)du.$$

Then we have the following result.

Theorem 5.23. $\left(X_t^f, \mathscr{F}_t\right)_{t \geqslant 0}$ is a $\mathbb{P}_i$-martingale for every $f \in C_0(E)$.

This is the martingale description for continuous-time Markov chains. Indeed, this description holds for more general Markov processes.

5.7   Supplements and Exercises

(1) Prove parts (1)-(3) of Lemma 5.2, where the mathematical expression
for (3) is: if $\left(X_t^{(i)}\right)_{n \geqslant 0}$, $i = 1, 2$ are submartingales, then so is the
maximal $\left(X_t^{(1)} \vee X_t^{(2)}\right)_{n \geqslant 0}$. Recall that $a \vee b = \max\{a, b\}$ and $a \wedge b = \min\{a, b\}$.

(2) Prove that $r\mathbb{P}\left[\sup_n X_n \geqslant r\right] \leqslant EX_0$ for every nonnegative supermartingale $X = (X_n)$ and every real number $r > 0$.

(3) If (X_t) is a martingale, then $(X_t \wedge c)$ is a supermartingale for every
constant c.

(4) Suppose that $(X_t)_{t \geqslant 0}$ is supermartingale and $\tau_0 \leqslant \tau_1 \leqslant \cdots$ are stopping times. Prove that $(X_{\tau_n})_{n \geqslant 0}$ is a supermartingale.

(5) Suppose that $(X_n)_{n \geqslant 1}$ are independent and identically distributed,
such that $\mathbb{E}X_1 = 0$ and $\mathbb{E}X_1^2 = 1$. Let $\mathscr{F}_n = \sigma(X_m : m \leqslant n)$
and $S_n = \sum_{m=1}^n X_m$. Then both $(S_n, \mathscr{F}_n)_{n \geqslant 1}$ and $(S_n^2 - n, \mathscr{F}_n)_{n \geqslant 1}$ are
martingales.

(6) If $(X_n)_{n \geqslant 0}$ is a nonnegative submartingale, then so is

$$X_n^* = \max_{i \leqslant n} X_i, \qquad n \geqslant 0.$$

(7) Suppose $(Y_n)_{n \geqslant 0}$ is a Markov chain on countable state space E, and
transition matrix $P = (p_{ij})$. Let $f = (f(i), i \in E)$ be nonnegative and
bounded, satisfying

$$f(i) = \sum_j p_{ij} f(j), \quad i \in E. \tag{5.6}$$

Set $X_n = f(Y_n)$ and $\mathscr{F}_n = \sigma(Y_m : m \leqslant n)$. Then $(X_n, \mathscr{F}_n)$ is a
martingale. If we replace (5.6) by

$$f(i) \geqslant \sum_j p_{ij} f(j), \quad i \geqslant 0,$$

then X_n is a supermartingale.

(8) Suppose $(Y_n)_{n \geqslant 0}$ is a Markov chain on countable state space E, and
transition matrix $P = (p_{ij})$. Let f be a non-trivial (nonzero) eigenvector of P, that is, there exists $\lambda > 0$ such that

$$\lambda f(i) = \sum_j p_{ij} f(j), \quad i \in E.$$

Define $X_n = \lambda^{-n} f(Y_n)$ and $\mathscr{F}_n = \sigma(Y_m : m \leqslant n)$. Prove that
$(X_n, \mathscr{F}_n)$ is a martingale.

(9) Suppose $(X_t)_{t \geqslant 0}$ is a Q-process on $\mathbb{Z}_+$. If $y = (y(i) : i \geqslant 0)$ is a nonzero solution to $Qy = \lambda y$, then $Y_t = e^{-\lambda t} y(X(t))$ is a martingale.

(10) Under the assumptions in Corollary 5.5, prove that $X_T \in \mathscr{F}_T$, $[T > S] \in \mathscr{F}_S$, and $X_T I_{[T \leqslant S]} \in \mathscr{F}_S$.

(11) Suppose that X_n is a symmetric random walk on $\mathbb{Z}$. Let

$$\tau = \min \{n : X_n \in \{N, -M\}\}, \quad N, M \geqslant 0$$

and

$$\mathscr{F}_n = \sigma(X_m, m \leqslant n).$$

Prove

(a) $(X_n, \mathscr{F}_n)$ is a martingale;

(b) $\tau \wedge n$ is a stopping time for every $n \geqslant 1$;

(c) $\mathbb{P}_0[X_\tau = N] = 1 - \mathbb{P}_0[X_\tau = -M] = \dfrac{M}{M + N}$.

Hint: Apply Doob's stopping theorem to stopping times 0 and τ.

(12) Suppose X_n is an asymmetric random walk on Z (that is, $p_{ii+1} = p$, $p_{ii-1} = q$, $p + q = 1$). Let

$$\tau = \min \{n : X_n \in \{N, -M\}\}, \quad N, M \geqslant 0$$

and

$$\mathscr{F}_n = \sigma(X_m, m \leqslant n).$$

Prove

$$\mathbb{P}_0[X_\tau = N] = 1 - \mathbb{P}_0[X_\tau = -M] = \frac{1 - (q/p)^M}{1 - (q/p)^{M+N}}.$$

Hint: Use X_n to construct a martingale and apply Doob's stopping theorem.

(13) Suppose $(Y_i)_{i \geqslant 0}$ are random variables and $\phi(x)$ is symmetric on $\mathbb{R}$, and increasing on $\mathbb{R}^+$. Assume $\phi(0) = 0$, and $(\phi(Y_i))_{0 \leqslant i \leqslant n}$ is a submartingale. Fix $0 = s_0 \leqslant s_1 \leqslant \cdots \leqslant s_n$. Prove

$$\mathbb{P}[|Y_i| \leqslant s_i, 1 \leqslant i \leqslant n] \geqslant 1 - \sum_{i=1}^{n} \frac{E[\phi(Y_i)] - E[\phi(Y_{i-1})]}{\phi(s_i)}.$$

(If $\phi(x) = |x|^p$, $p \geqslant 1, s_1 = s_2 = \cdots = \lambda$, then Kolmogorov's inequality follows.)

(14) Suppose $(X_t, \mathscr{F}_t)_{t \geqslant 0}$ is a right-continuous martingale, and f is a nonnegative convex function. Then for every $\epsilon > 0$,

$$\mathbb{P}\left[\sup_{t \geqslant 0} f(X_t) \geqslant \epsilon\right] \leqslant \frac{1}{\epsilon} \sup_{t \geqslant 0} E f(X_t).$$

(15) Suppose $(X_n, \mathscr{F}_n)_{n \geq 0}$ is a martingale. Prove

(a) $\mathbb{E}X_n = \mathbb{E}X_0$ for every $n \geq 0$;

(b) $\mathbb{E}[X_{n+m}|\mathscr{F}_n] = X_n$ for every $n, m \geq 0$;

(c) by letting $Y_n = X_{2n}$ and $\mathscr{G}_n = \mathscr{F}_{2n}$, $(Y_n, \mathscr{G}_n)$ is a martingale.

(16) (Kolmogorov's 0-1 law) Suppose $(X_n)_{n \geq 0}$ is a sequence of independent random variables and $\mathscr{T}$ is the tail σ-algebra, that is, $\mathscr{T} = \bigcap_{n \geq 1} \mathscr{H}_n$, $\mathscr{H}_n = \sigma(X_n, X_{n+1}, \cdots)$. Prove that either $\mathbb{P}(A) = 1$ or $\mathbb{P}(A) = 0$ for every $A \in \mathscr{T}$. Thus for any random variable $Y \in \mathscr{T}$, there exists a constant $c : -\infty \leq c \leq \infty$ such that $\mathbb{P}[Y = c] = 1$.

(17) Suppose $X, X_n, n \geq 1$ are integrable random variables, and when $\alpha \to \infty$, $\mathbb{P}[|X_n| \geq \alpha] \to 0$ uniformly in $n \geq 1$. Then

$$\lim_{\alpha \to \infty} \sup_n \int_{[|X_n| \geq \alpha]} |X| d\mathbb{P} = 0.$$

(18) Construct an example such that the martingale (X_t) is bounded in L^1 but not uniformly integrable.

(19) Suppose $\left\{ (X_t^{(n)})_{t \geq 0}, n \geq 1 \right\}$ is a sequence of $(\mathscr{F}_t)$-martingales. Assume $\lim_{n \to \infty} X_t^{(n)} = X_t$ in L^p $(p \in [1, \infty])$ for each $t \geq 0$. Then $(X_t)_{t \geq 0}$ is also an $(\mathscr{F}_t)$-martingale.

(20) Suppose $X_n, Y_n, n \geq 0$ are $(\mathscr{F}_n)_{n \geq 0}$-martingales and T is a stopping time. Suppose further $X_T = Y_T$ on $[T < \infty]$. Then

$$Z_n = \begin{cases} X_n, & n < T; \\ Y_n, & n \geq T \end{cases}$$

is a martingale.

(21) Consider the Q-process (X_t) has the infinitesimal generator Ω. Let $\tau_n, n \geq 0$ be the nth jump epoch and $\tau_\infty = \lim_{n \to \infty} \tau_n$, satisfying $\mathbb{P}[\tau_n < \infty] = 1$ and $\mathbb{P}[\tau_\infty = \infty] = 1$. Let $\mathscr{F}_t = \sigma\{X_s : s \leq t\}$. Assume for every $i, f(i) > 0, \sum_{j \neq i} q_{ij} f(j) < \infty$. Define

$$Z(t) = f(X_t) \exp\left[-\int_0^t \left(\frac{\Omega f}{f} \right)(X_s) ds \right], \quad Z_n(t) = Z(t \wedge \tau_n), \ n \geq 0.$$

Then

(a) $\mathbb{E}_i Z_n(t) = f(i)$, $t \geq 0$;

(b) $(Z(\tau_n), \mathscr{F}_{\tau_n}, \mathbb{P}_i)$ is a martingale and $(Z(t), \mathscr{F}_t, \mathbb{P}_i)$ is a supermartingale for each i.

Chapter 6

Brownian Motion

6.1 Brownian Motion

From the view of either practice or mathematical theory, Brownian motion is an extremely important stochastic process. In 1827, British botanist Robert Brown found this motion of pollen grains in water, which is caused by collisions from the molecules of water. This motion produces a kind of random force. The collision number caused by the pollens is very large, up to about 10^{21}. Therefore, the motion of pollen grains can be viewed as a stochastic motion caused by a large amount of tiny force. Let B_t be the position of a pollen grain at time t. If the liquid is even spacing, then the displacement $B_{t_2} - B_{t_1}$ from time t_1 to time t_2 is the sum of many nearly independent tiny displacements, that is, the sum of many tiny independent random variables. According to the central limit theorem, we have

$B_{t_2} - B_{t_1} \sim \mathcal{N}(a(t_1, t_2), \sigma(t_1, t_2))$,   (normal property of the increments)

where $\mathcal{N}(a, \sigma)$ denotes the normal distribution with mean a and variance σ^2. By the even spacing of liquid,

$a(t_1, t_2) = 0$,

$\sigma(t_1, t_2)^2 = \sigma^2(t_2 - t_1)$ (the spread is proportional to time).

Here $\sigma > 0$ is the medium constant (depending on the property of liquid), independent of time and space. Further, since $B_{t_n} - B_{t_{n-1}}, \cdots, B_{t_2} - B_{t_1}$ are the sum of many tiny almost independent displacements, they are assumed to be independent.

Definition 6.1. The real-valued process $(B_t)_{t \geqslant 0}$ is called **Brownian motion (BM)** or **Wiener process**, if it satisfies the following two conditions.

(1) *Independent increment property.* For every $n \geqslant 2$ and $0 = t_0 < t_1 < t_2 < \cdots < t_n$, the increments

$$B_{t_1} - B_{t_0}, \; B_{t_2} - B_{t_1}, \; \cdots, \; B_{t_n} - B_{t_{n-1}}$$

are mutually independent (the process is called **process with independent increment**), and these increments are also independent of B_{t_0}.

(2) *Normal property.* For every $s < t$, $B_t - B_s \sim \mathcal{N}(0, \sigma|t-s|^{1/2})$, where $\sigma > 0$ is a constant.

Let $(X_t)_{t \geqslant 0}$ be a process with independent increments. If X_{t_0} is independent of $X_{t_1} - X_{t_0}, \cdots, X_{t_n} - X_{t_{n-1}}$ ($t_0 = 0$, $t_n = t$), then $(X_t)_{t \geqslant 0}$ is a Markov process. Indeed, for any $0 = t_0 < t_1 < \cdots < t_n$, let

$$Y_0 = X_{t_0}, \ Y_1 = X_{t_1} - X_{t_0}, \ \cdots, \ Y_n = X_{t_n} - X_{t_{n-1}}.$$

Then $(Y_n)_{n \geqslant 0}$ are independent random variables, and $X_{t_n} = \sum_{i=0}^{n} Y_i$ is a sum of independent random variables, so it is Markovian:

$$\mathbb{P}[X_{t_n} \leqslant \alpha | X_{t_0}, \cdots, X_{t_{n-1}}] = \mathbb{P}[X_{t_n} \leqslant \alpha | X_{t_{n-1}}].$$

For BM, if assume that $\mathbb{P}[B_0 = 0] = 1$, then $(B_t)_{t \geqslant 0}$ is a Markov process. As seen above, if we denote $s = t_{n-1}$, then

$$\mathbb{P}[B_t \leqslant \alpha | B_{t_0}, \cdots, B_{t_{n-1}}] = \mathbb{P}[B_t \leqslant \alpha | B_s].$$

On the other hand, by the virtue of property of independent increments, we have

$$\mathbb{E}[f(B_s)g(B_t)] = \mathbb{E}[f(B_s - B_0)g(B_t - B_s + B_s - B_0)]$$

$$= \frac{1}{2\pi\sigma^2\sqrt{s(t-s)}} \int_{\mathbb{R}^2} \exp\left[-\frac{x^2}{2\sigma^2 s}\right] \exp\left[-\frac{y^2}{2\sigma^2(t-s)}\right] f(x)g(x+y)\mathrm{d}x\mathrm{d}y$$

$$= \frac{1}{2\pi\sigma^2\sqrt{s(t-s)}} \int_{\mathbb{R}} \exp\left[-\frac{x^2}{2\sigma^2 s}\right] f(x)\mathrm{d}x \int_{\mathbb{R}} \exp\left[-\frac{y^2}{2\sigma^2(t-s)}\right] g(x+y)\mathrm{d}y$$

$$= \frac{1}{2\pi\sigma^2\sqrt{s(t-s)}} \int_{\mathbb{R}} \exp\left[-\frac{x^2}{2\sigma^2 s}\right] f(x)\mathrm{d}x \int_{\mathbb{R}} \exp\left[-\frac{(y-x)^2}{2\sigma^2(t-s)}\right] g(y)\mathrm{d}y$$

$$= \int_{\mathbb{R}} p(s, 0, \mathrm{d}x)f(x) \int p(t-s, x, \mathrm{d}y)g(y),$$

where

$$p(t, x, \mathrm{d}y) = \frac{1}{\sqrt{2\pi t}\,\sigma} \exp\left[-\frac{(y-x)^2}{2\sigma^2 t}\right] \mathrm{d}y.$$

We have a similar expression for $\prod_{k=1}^{m} f_k(B_{t_k})$. This shows that $(B_t)_{t \geqslant 0}$ is a Markov process with transition probability $p(t, x, \mathrm{d}y)$.

6.2   The Trajectory Property

We begin this section with a fundamental result on the trajectory property of stochastic processes. Its proof can be found in many textbooks, e.g. [44; §15.1] or [66; §3.2].

Theorem 6.2 (Kolmogorov's criterion). Suppose there exist $\alpha, \varepsilon > 0$ such that

$$\mathbb{E}[|X_t - X_{t+h}|^{\alpha}] \leqslant \text{const.} |h|^{1+\varepsilon}, \qquad t,\ t+h \in [a,b].$$

Then the process $(X_t)_{t \in [a,b]}$ is a.s. continuous.

For BM, we have

$$\mathbb{E}(|B_t - B_{t+h}|^{2m}) = \frac{1}{\sqrt{2\pi |h|}\,\sigma} \int_{-\infty}^{+\infty} y^{2m} \exp\left[-\frac{y^2}{2\sigma^2 |h|}\right] dy$$

$$= \left(\sigma\sqrt{|h|}\right)^{2m} \frac{1}{\sqrt{2\pi}} \int_{-\infty}^{+\infty} z^{2m} \exp\left[-\frac{z^2}{2}\right] dz$$

$$= C_m |h|^m, \qquad m \in \mathbb{N}.$$

From this we see that the trajectories of (B_t) are a.s. continuous. Applying the separable property, we can construct a separable version, whose trajectories are all continuous. On the other hand,

$$P(t)f(x) := \int p(t,x,\mathrm{d}y)f(y) = \frac{1}{\sqrt{2\pi t}\,\sigma} \int_{-\infty}^{+\infty} \exp\left[-\frac{(y-x)^2}{2t\sigma^2}\right] f(y)\mathrm{d}y$$

$$\implies P(t)C_b(\mathbb{R}) \subset C_b(\mathbb{R})$$

$$\implies (P(t))_{t \geqslant 0} \text{ is Feller.}$$

Therefore, (B_t) is a strong Markov process. Indeed, we have a stronger result:

Theorem 6.3. $(B_t)_{t \geqslant 0}$ is a strong Markov process with respect to $(\mathscr{F}_{t+})_{t \geqslant 0}$.

As the constant σ is often inconvenient, we use the following convention.

Definition 6.4. (B_t) is called a standard BM, if it is a BM with $\sigma = 1$, and all trajectories are continuous with $B_0 \equiv 0$.

After studying several simple properties of BM, we now study some "marvelous" properties.

Theorem 6.5. Assume $(B_t)_{t \geqslant 0}$ is a BM. Then each of the following processes is also a BM.

(1) $(-B_t)_{t \geqslant 0}$ (*symmetry by reflect*);
(2) $(B_{t+s} - B_s)_{t \geqslant 0}$, $s \geqslant 0$ fixed (*transform of the original*);

(3) $(cB_{t/c^2})_{t \geqslant 0}$, $c > 0$ fixed (*scaling transform*);

(4) $(tB_{1/t})_{t \geqslant 0}$ with convention that it is zero at $t = 0$ (*time reciprocal*);

(5) Fix $u > 0$, $(B_u - B_{u-t})_{0 \leqslant t \leqslant u}$ (*time reverse*).

Proof We prove only (4). Let $X_t = tB_{1/t}$, $t > 0$.

a) Since $B_t \sim \mathcal{N}(0, \sqrt{t})$, by the scaling property of the normal distribution, we have

$$X_t \sim \mathcal{N}\left(0, \sqrt{t^2 \cdot 1/t}\right) = \mathcal{N}(0, \sqrt{t}).$$

b) Moreover, if $s \leqslant t$, then

$$\mathbb{E}[B_s B_t] = \mathbb{E}[B_s(B_t - B_s)] + \mathbb{E}B_s^2 = \mathbb{E}B_s\,\mathbb{E}(B_t - B_s) + \mathbb{E}B_s^2 = \mathbb{E}B_s^2 = s.$$

Then $\mathbb{E}[B_s B_t] = s \wedge t$, and

$$\mathbb{E}[X_s X_t] = st\mathbb{E}[B_{1/s}B_{1/t}] = s \wedge t.$$

Therefore, if $0 < t_1 < t_2$, then $\mathbb{E}(X_{t_2} - X_{t_1})^2 = t_2 - 2t_1 \wedge t_2 + t_1 = t_2 - t_1$. Thus $X_{t_2} - X_{t_1} \sim \mathcal{N}(0, \sqrt{t_2 - t_1})$.

c) On the other hand, for $0 < t_1 < t_2 < t_3 < t_4$, by the normal property of the increments we have

$$\mathbb{E}[(X_{t_2} - X_{t_1})(X_{t_4} - X_{t_3})] = t_2 \wedge t_4 - t_1 \wedge t_4 - t_2 \wedge t_3 + t_1 \wedge t_3$$
$$= t_2 - t_1 - t_2 + t_1 = 0.$$

This prove that the process has independent increments (it is well-known that the normal process is irrelative whenever independent). But for any $0 < t_0 < t_1 < \cdots < t_n$ and $\alpha_0, \cdots, \alpha_n \in \mathbb{R}$,

$$\sum_{i=0}^{n} \alpha_i X_{t_i} = \sum_{i=0}^{n} \alpha_i t_i B_{1/t_i}$$

is normal. Therefore $(X_t)_{t > 0}$ is a normal process with independent increments.

d) Finally, every trajectory of (X_t) is continuous in $(0, \infty)$. As for $t = 0$, according to the property of BM, we can choose a separable subset D of R_+ with 0 the limit such that

$$\lim_{t \in D,\, t \downarrow 0} X_t = 0, \qquad \text{a.s.}$$

Thus, we define additionally that $X_0 \equiv 0$ and make (X_t) (right-)continuous at $t = 0$. $\square$

Remark 6.6. Part (4) of Theorem 6.5 leads to the following strong law of large number:

$$\mathbb{P}\left[\lim_{t \uparrow \infty} \frac{1}{t} B_t = 0 \right] = 1.$$

Theorem 6.7. With probability 1, BM is not Lipschitz continuous at any $t \geqslant 0$. Therefore, BM is nowhere differential.

Proof If $B_t(\omega)$ is Lipschitz continuous at some $t \in [0,1)$ then there exist $\ell, m \in \mathbb{N}$ such that the following estimates of increments

$$\left| B_{\frac{k+1}{n}} - B_{\frac{k}{n}} \right| \leqslant \frac{\ell}{n}$$

hold for all $n \geqslant m$ and three successive k's in $\{0, \cdots, n-1\}$. Denote this event by Λ. Set

$$\Lambda_{\ell,m} = \bigcap_{n=m}^{\infty} \bigcup_{k=0}^{n-1} \bigcap_{j=0}^{2} \left[\left| B_{\frac{k+j+1}{n}} - B_{\frac{k+j}{n}} \right| \leqslant \frac{\ell}{n} \right].$$

Then $\Lambda = \bigcup_{\ell=1}^{\infty} \bigcup_{m=1}^{\infty} \Lambda_{\ell,m}$. We will prove $\mathbb{P}(\Lambda) = 0$, for which we need only prove that $\mathbb{P}(\Lambda_{\ell,m}) = 0$. Indeed,

$$\mathbb{P}(\Lambda_{\ell,m}) \leqslant \varlimsup_{n\to\infty} \mathbb{P}\left\{ \bigcup_{k=0}^{n-1} \bigcap_{j=0}^{2} \left[\left| B_{\frac{k+j+1}{n}} - B_{\frac{k+j}{n}} \right| \leqslant \frac{\ell}{n} \right] \right\}$$

$$\leqslant \varlimsup_{n\to\infty} n\mathbb{P}\left\{ \bigcap_{j=0}^{2} \left[\left| B_{\frac{j+1}{n}} - B_{\frac{j}{n}} \right| \leqslant \frac{\ell}{n} \right] \right\} \quad \text{(by stationary property)}$$

$$= \varlimsup_{n\to\infty} n \left(\mathbb{P}\left[\left| B_{\frac{1}{n}} \right| \leqslant \frac{\ell}{n} \right] \right)^3 \quad \text{(by independence of increments)}$$

$$= \varlimsup_{n\to\infty} n \left(\mathbb{P}\left[|B_1| \leqslant \frac{\ell}{\sqrt{n}} \right] \right)^3 \quad \text{(by scaling transform)}$$

$$\leqslant \text{const.} \varlimsup_{n\to\infty} n \left(\frac{1}{\sqrt{n}} \right)^3 \quad \left(\text{since } \int_0^z e^{-x^2/2}\mathrm{d}x \sim z \text{ as } z \to 0 \right)$$

$$= 0.$$

The last step indicates why we choose three successive intervals. □

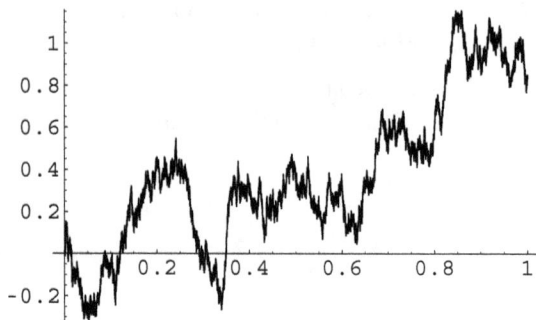

Sample trajectory of BM

Since the finite variance function is differential almost everywhere, we have

Corollary 6.8. BM has unbounded variance in every finite interval.

The following law of iterated logarithm is due to Lévy, and has many extension to the sequence of independent and identically distributed random variables. See, for example, [52].

Theorem 6.9 (Law of iterated logarithm for BM (Lévy)).

$$\mathbb{P}\left[\overline{\lim_{\delta \downarrow 0}} \sup_{\substack{0 \leqslant s \leqslant t \leqslant 1 \\ t-s < \delta}} \frac{|B_t - B_s|}{(2\delta \log \log \frac{1}{\delta})^{1/2}} = 1 \right] = 1.$$

6.3   Martingale Property of Brownian Motion

Suppose $(B_t)_{t \geqslant 0}$ is the standard Brownian motion defined on a probability space $(\Omega, \mathscr{F}, \mathbb{P})$. Let $\mathscr{F}_t^0 = \sigma(B_u : u \leqslant t)$ and $\mathscr{F}_t = \sigma(\mathscr{F}_t^0 \cup \mathscr{N})$, where $\mathscr{N}$ be the total of all $\mathbb{P}$-null sets. We call $(\mathscr{F}_t)$ is the **completion** of $(\mathscr{F}_t^0)$. The reason for making the completion is to make $T_A = \inf\{t \geqslant 0 : B_t \in A\}$ to be measurable for every $A \in \mathscr{B}(\mathbb{R})$.

Theorem 6.10. With respect to $(\mathscr{F}_t)_{t \geqslant 0}$ and $\mathbb{P}_x$ $(x \in \mathbb{R})$, the following assertions hold.

(1) $(B_t)_{t \geqslant 0}$ is a martingale.

(2) $(B_t^2 - t)_{t \geqslant 0}$ is a martingale.

(3) For evert $\theta \in \mathbb{R}$, $\left(\exp\left[\theta B_t - \frac{1}{2} t \theta^2 \right] \right)_{t \geqslant 0}$ is a martingale, which is called the **exponential martingale of BM** (the Laplace transform of BM).

(4) If $f \in C^2(\mathbb{R})$ (the set of all functions having continuous second order derivative) such that f and $\Delta f = \dfrac{\mathrm{d}^2 f}{\mathrm{d}x^2}$ are bounded, then

$$\left(f(B_t) - \int_0^t \frac{1}{2}\Delta f(B_u)\mathrm{d}u \right)_{t \geqslant 0}$$

is a martingale.

Proof (1) By the Markov property,

$$\mathbb{E}_x[f(B_t)|\mathscr{F}_s] = \mathbb{E}_{B_s} f(B_{t-s}), \qquad f \in {}_b\mathscr{B}(\mathbb{R}).$$

Then by the monotone class theorem, the equation holds for each $f \in \mathscr{B}(\mathbb{R})$ such that $f(B_t)$ is integrable. In particular, it follows from the integrability of B_t that

$$\mathbb{E}_x[B_t|\mathscr{F}_s] = \mathbb{E}_{B_s}(B_{t-s}).$$

But

$$\mathbb{E}_x B_t = \int p(t,x,\mathrm{d}y)y = \frac{1}{\sqrt{2\pi t}} \int_{-\infty}^{+\infty} \exp\left[-\frac{(y-x)^2}{2t} \right] y\mathrm{d}y$$

$$= \frac{1}{\sqrt{2\pi t}} \int_{-\infty}^{+\infty} \exp\left[-\frac{(y-x)^2}{2t} \right][(y-x) + x]\mathrm{d}y$$

$$= x \cdot \frac{1}{\sqrt{2\pi t}} \int_{-\infty}^{+\infty} \exp\left[-\frac{(y-x)^2}{2t} \right]\mathrm{d}y$$

$$= x.$$

Thus $\mathbb{E}_x[B_t|\mathscr{F}_s] = B_s$.

(2) $\mathbb{E}_x[B_t^2|\mathscr{F}_s] = \mathbb{E}_x[(B_t - B_s)^2 + 2B_s(B_t - B_s) + B_s^2|\mathscr{F}_s] = (t - s) + 0 + B_s^2$.

(3) Denote

$$X_t = \exp\left[\theta B_t - \frac{\theta^2}{2}t \right], \qquad t \geqslant 0,$$

$$X_t^s = \exp\left[\theta(B_t - B_s) - \frac{\theta^2}{2}(t - s) \right], \qquad t \geqslant s.$$

Since $B_t - B_s \sim \mathcal{N}(0, \sqrt{t-s})$, we have $\mathbb{E}X_t^s = 1$. On the other hand, note that X_t^s and $\mathscr{F}_s$ are independent, and $X_t = X_s X_t^s$. Therefore

$$\mathbb{E}_x[X_t|\mathscr{F}_s] = \mathbb{E}_x[X_s X_t^s|\mathscr{F}_s] = X_s \mathbb{E}X_t^s = X_s.$$

(4) The proof for this assertion is postponed to Section 8.5 in Chapter 8.

$\square$

6.4   Multi-Dimensional Brownian Motion

Definition 6.11. Let $B_t^1, \cdots, B_t^d$ be mutually independent (standard) BM. Then $B_t = (B_t^1, \cdots, B_t^d)$ is called the d-dimensional Brownian motion, denoted by BM^d.

Like one-dimensional BM, BM^d has the transition probability function:

$$p(t, x, \mathrm{d}y) = \frac{1}{(2\pi t)^{d/2}} \exp\left[-\frac{1}{2t} \sum_{i=1}^{d} (y_i - x_i)^2 \right] \mathrm{d}y, \quad x, y \in \mathbb{R}^d.$$

It is easy to prove that all assertions in Theorem 6.5 hold for BM^d.

Theorem 6.12. With respect to $(\mathscr{F}_t)_{t \geqslant 0}$ and $\mathbb{P}_x$ $(x \in \mathbb{R}^d)$, the following assertions for BM^d (B_t) hold.

(1) $(B_t)_{t \geqslant 0}$ is a martingale.

(2) $(|B_t|^2 - dt)_{t \geqslant 0}$ is a martingale, where $|x| = (x_1^2 + \cdots + x_d^2)^{1/2}$, $x \in \mathbb{R}^d$.

(3) For every $\theta \in \mathbb{R}^d$, $\left(\exp\left[\langle \theta, B_t \rangle - \frac{1}{2} t |\theta|^2 \right] \right)_{t \geqslant 0}$ is a martingale, which is called **exponential martingale for BM^d** (the Laplace transform of BM^d), where $\langle x, y \rangle = \sum_{i=1}^{d} x_i y_i$, $x, y \in \mathbb{R}^d$.

(4) If $f \in C^2(\mathbb{R}^d)$ such that f and $\Delta f = \sum_{i=1}^{d} \dfrac{\partial^2 f}{\partial x_i^2}$ are bounded, then

$$\left(f(B_t) - \int_0^t \frac{1}{2} \Delta f(B_u) \mathrm{d}u \right)_{t \geqslant 0}$$

is a martingale.

6.5   Supplements and Exercises

(1) (a) Prove that (X, Y) has the two-dimensional normal distribution if and only if for any $\alpha, \beta \in \mathbb{R}$, $\alpha X + \beta Y$ has the normal distribution.

   (b) Try to get an example that each of X and Y has the normal distribution and that are irrelative, but X, Y are not independent.

(2) Let $T(\omega) = \{0 \leqslant t \leqslant 1 : B_t(\omega) = 0\}$. Prove that for every $\omega \in \Omega$, $T(\omega)$ is a closed set, and has null Lebesgue measure.

(3) Use the law of iterated logarithm for BM:

$$\limsup_{t \downarrow 0} \frac{B_t}{\sqrt{2t \log \log t^{-1}}} = 1, \qquad \liminf_{t \downarrow 0} \frac{B_t}{\sqrt{2t \log \log t^{-1}}} = -1,$$

to prove $T(\omega)$ in the previous item is an infinite set, and is a complete set (has no isolated point).

(4) Let B_t be the standard BM, and $\tau_x = \inf\{t \geqslant 0 : B_t = x\}$, $x \in \mathbb{R}$. Prove that for the (standard) BM, the density function for τ_x is

$$(2\pi t^3)^{-1/2} \exp[-x^2/2t].$$

Hint: By Property (3) in Theorem 6.10,

$$\mathbb{E}\left[\exp\left(\theta x - \frac{1}{2}\theta^2 \tau_x \right) : \tau_x < \infty \right] = 1.$$

From which, we can get $\mathbb{P}[\tau_x < \infty] = 1$. Next, set $\lambda = \frac{1}{2}\theta^2$, we have the Laplace transform of τ_x is

$$\mathbb{E}\exp[-\lambda \tau_x] = \exp[-x(2\lambda)^{1/2}], \qquad \forall \lambda > 0.$$

This gives us the conclusion.

(5) Prove the probability transition density function

$$p(t,x,y) = \frac{1}{\sqrt{2\pi t}} \exp[-(x-y)^2/2t], \quad t \geqslant 0, \ x,y \in \mathbb{R}$$

satisfies

$$\frac{\partial p}{\partial t} = \frac{1}{2}\frac{\partial^2 p}{\partial x^2}.$$

(6) Prove

$$\sum_{k=1}^{2^n} \left[B_{kt/2^n} - B_{(k-1)t/2^n} \right]^2 \xrightarrow[L^2]{a.s.} t, \quad n \to \infty$$

for each $t > 0$. Thus BM does not have the bounded variation in any finite interval.

(7) Prove the following Kolmogorov's inequality for BM:

$$\mathbb{P}\left[\sup_{0 \leqslant s \leqslant t} |B_s| > r \right] \leqslant \frac{t}{r^2}.$$

(8) Prove the following Doob's L^2-inequality:

$$\mathbb{E}\left(\max_{0 \leqslant s \leqslant t} |B_s|^2 \right) \leqslant 4\mathbb{E}B_t^2.$$

(9) Let $W_t = \int_0^t B_s ds$.
  (a) Prove $E[W_t] = 0$ and $E[W_t^2] = t^3/3$.
  (b) Find the conditional distribution of W_t under $B_t = x$.
  (c) Prove $(W_t - tB_t)$ is a martingale.

(10) Prove that $B_t^3 - 3tB_t, B_t^4 - 6tB_t^2 + 3t^2, \cdots$ are martingales.

(11) Let (B_t) be a BM. Prove

(a) $\mathbb{E}\exp\left[\lambda \int_0^t B_s \mathrm{d}s\right] = \exp(\lambda^2 t^3/6)$;

(b) $\mathbb{E}\exp\left[\lambda \int_0^t sB_s \mathrm{d}s\right] = \exp(\lambda^2 t^5/15)$.

(12) Let (B_t) be a BM. Set $X_t = e^{-t}B_{e^{2t}}$ for $t \geqslant 0$ and $Y_t = B_t - tB_1$ for $t \in [0,1]$. Compute $\mathbb{E}X_tX_s$ for $t, s \geqslant 0$ and $\mathbb{E}Y_tY_s$ for $s, t \in [0,1]$.

(13) Let $M_t = \max\limits_{0\leqslant s\leqslant t} B_s$. Prove

(a) $Y_t = M_t - B_t$ is a Markov process;

(b) The processes (Y_t) and $(|B_t|)$ are equivalent (that is they have the same finite-dimensional distributions).

(14) Consider a real-valued integrable and continuous function f such that

$$\int_{-\infty}^{\infty} f(x)\mathrm{d}x = a > 0.$$

Define the process

$$Y_t = \frac{1}{\sqrt{t}} \int_0^t f(B_s)\mathrm{d}s.$$

Prove the limits $\lim\limits_{t\to\infty} E[Y_t]$ and $\lim\limits_{t\to\infty} E[Y_t^2]$ exist, and find these limits.

(15) Prove parts (1)-(3) of Theorem 6.5.

(16) Let (B_t) be a BM and $\mathscr{F}_t = \sigma(B_s, s \leqslant t)$. Let τ be a stopping time with respect to $\mathscr{F}_{t+} := \bigcap\limits_{s>t} \mathscr{F}_s$ and let $\tilde{B}_t = B_{t+\tau} - B_\tau$. Then $(\tilde{B}_t)$ is still a BM.

Hint: Use Theorem 6.3 and the characteristic function to prove $(\tilde{B}_{t_1}, \cdots, \tilde{B}_{t_k})$ and $(B_{t_1}, \cdots, B_{t_k})$ have the same distribution for any $t_1 < t_2 < \cdots < t_k$.

(17) Let (B_t) be a standard BM and $\tau = \inf\{t : B_t \notin (a,b)\}$ for $a < 0 < b$. Use the Doob's stopping theorem for martingales to prove

$$\mathbb{P}[B_\tau = b] = 1 - \mathbb{P}[B_\tau = a] = -a/(b-a).$$

(18) Prove the properties stated in Theorem 6.12.

(19) Let B_t be a BM and $X_t = e^{B_t}$ be its geometric BM. Compute its diffusion coefficient

$$a(x) = \lim_{h\downarrow 0} \frac{\mathbb{E}[(X_{t+h} - X_t)^2|X_t = x]}{h}, \qquad 0 < x < \infty,$$

and its drift coefficient

$$b(x) = \lim_{h\downarrow 0} \frac{\mathbb{E}[X_{t+h} - X_t|X_t = x]}{h}, \qquad 0 < x < \infty.$$

(20) Let B_t be a (standard) d-dimensional BM^d, and $R_t = |B_t|$ be its radial BM.

(a) Prove R_t is a Markov process, and its transition probability function is

$$p(t, x, \mathrm{d}y) = t^{-1} \exp\left[-\frac{x^2 + y^2}{2t} \right] (xy)^{1-d/2} I_{d/2-1}\left(\frac{xy}{t} \right) y^{d-1} \mathrm{d}y,$$

where I_ν is the Bessel function

$$I_\nu(z) = \sum_{k=0}^{\infty} \frac{(z/2)^{2k+\nu}}{k!\Gamma(k+\nu+1)}.$$

(b) Let $T = \inf\{t \geqslant 0 : R_t = r\}$. Use Doob's stopping theorem for martingale to prove $\mathbb{E}T = r^2/d$.

Chapter 7

Stochastic Integral and Diffusion Processes

7.1 Stochastic Integral

The purpose of this chapter is to define the following integral:

$$\int_0^\infty \phi_t \mathrm{d}B_t, \tag{7.1}$$

where B_t is the Brownian motion, $\phi = (\phi_t)$ is an $(\mathscr{F}_t)$-adapted process, that is $\phi_t \in \mathscr{F}_t := \sigma\{B_s : s \leqslant t\}$. Recall that, to define the Riemannian integral $\int_0^1 \psi_t \mathrm{d}t$ for function ψ, we use the Riemannian sum

$$\sum_{i=0}^n \psi_{\theta_i}(t_{i+1} - t_i)$$

as an approximation. Here $0 = t_0 < t_1 < \cdots < t_{n+1} = 1$, $\theta_i \in [t_i, t_{i+1}]$. If ψ is random, $\psi(\theta_i)$ is understood as the observed value at time t. Although the above approximation can be defined point-wise (for each ω), it is too tough, as the random errors is not taken into account. The finer approximation must be as following

$$\sum_{i=0}^n [\psi_{\theta_i}(t_{i+1} - t_i) + \phi_{\theta_i}(B_{t_{i+1}} - B_{t_i})]$$

where $(B_{t_{i+1}} - B_{t_i})_{i \geqslant 1}$ are independent and normal distributed random variables with zero means. This is the usual error form. From this, we need to study the following integral

$$\int_0^1 \psi_t \mathrm{d}t + \int_0^1 \phi_t \mathrm{d}B_t.$$

The latter integral would be the stochastic integral with respect to the Brownian motion, considered in this chapter. Since the trajectories of Brownian motion (that is $(B_t(\omega))_{t \geqslant 0}$ for a given ω) have unbounded variation, the above integral (7.1) can not be viewed as the usual Lebesgue-Stieltjes integral (for fixed ω). To overcome the difficulty, many efforts were made in the history. For example, N. Wiener defined an integral in terms of integration by parts formula

$$\int_0^1 \phi_t dB_t = \psi_t B_t|_0^1 - \int_0^1 B_t \phi_t' dt$$

(cf. Exercise 2 in this chapter). However, this definition requires ϕ_t to be absolutely continuous, which excludes some important situation, such as $\int_0^1 B_t dB_t$. Thus we need to define (7.1) in a new convergence sense. The key point is that although the Brownian motion has unbounded variation, the quadratic variation is bounded. See Exercise 6 in Chapter 6 for the quadratic variation of BM. Therefore we can define the stochastic integral in the sense of L^2-isometric. This definition is not point-wise, but in the almost sure sense.

Let

$$\mathscr{L}^2 = \left\{ \phi = (\phi_t) : \phi \text{ is an adapted process such that } \mathbb{E} \int_0^\infty \phi_t^2 dt < \infty \right\}.$$

Define $||\phi||^2 = \mathbb{E} \int_0^\infty \phi_t^2 dt$ for $\phi \in \mathscr{L}^2$. Then it is easy to check that $|| \cdot ||$ is an L^2-norm on $\mathscr{L}^2$ (in the sense of equivalent class). Here, we deal with the infinite interval $[0, \infty)$, which includes the case of finite interval.

As for the usual definition of integral, we define the integral for "step-like" adapted processes, and then use an approximation procedure to extend the integrals to the whole $\mathscr{L}^2$.

Assume that $0 = t_0 < t_1 < \cdots < t_n < \infty$. Let $\xi_1, \cdots, \xi_n$ be random variables such that $\xi_i \in \mathscr{F}_{t_i}$. Notice that here for the measurability of ξ_i, we use t_i rather than t_{i+1}, since they have essential difference. For this, see Exercises 3 and 4 of this chapter. Let

$$\phi_t = \sum_{i=0}^{n-1} \xi_i I_{(t_i, t_{i+1}]}(t), \quad t \geqslant 0. \tag{7.2}$$

Then the stochastic integral of $\phi = (\phi_t)$ is defined as

$$I_\phi = \int_0^\infty \phi_t dB_t = \sum_{i=0}^{n-1} \xi_i (B_{t_{i+1}} - B_{t_i}).$$

Lemma 7.1. We have

$$\mathbb{E} I_\phi^2 = ||\phi||^2. \tag{7.3}$$

For general $\phi \in \mathscr{L}^2$, to define its stochastic integral I_ϕ, we need the following approximation result.

Proposition 7.2. For every $\phi \in \mathscr{L}^2$, there exists a sequence of the "step-like" functions $\phi^{(n)}$ as in (7.2) such that $\lim\limits_{n\to\infty} ||\phi^{(n)} - \phi|| = 0$.

By virtue of Proposition 7.2, we can define

Definition 7.3. For $\phi = (\phi_t) \in \mathscr{L}^2$, choose a sequence of "step-like" functions $\phi^{(n)}$ as in (7.2) such that $\lim\limits_{n\to\infty} ||\phi^{(n)} - \phi|| = 0$. Then we define I_ϕ as the limit of $I_{\phi^{(n)}}$ in $L^2(\mathbb{P})$. Namely,

$$\lim_{n\to\infty} \mathbb{E}\left(I_\phi - I_{\phi^{(n)}} \right)^2 = 0.$$

Obviously, the stochastic integral I_ϕ in Definition 7.3 is unique a.s., and is independent of the choice of "step-like" functions $\phi^{(n)}$.

Next, we turn back to prove Lemma 7.1 and Proposition 7.2.

Proof of Lemma 7.1. From (7.2) and the independent increment property of BM,

$$\mathbb{E}I_\phi^2 = \mathbb{E}\left[\sum_{i=0}^{n-1} \xi_i (B_{t_{i+1}} - B_{t_i}) \right]^2$$

$$= \mathbb{E}\left[\sum_{i=0}^{n-1} \xi_i^2 (B_{t_{i+1}} - B_{t_i})^2 + \sum_{i \neq j} \xi_i \xi_j (B_{t_{i+1}} - B_{t_i})(B_{t_{j+1}} - B_{t_j}) \right]$$

$$= \sum_{i=0}^{n-1} \mathbb{E}\xi_i^2 (t_{i+1} - t_i) = ||\phi||^2. \quad \square$$

Proof of Proposition 7.2. Without loss of generality, assume ϕ is bounded, otherwise let $\phi^{(n)} = \phi \wedge n \vee (-n)$. It is easy to see

$$\left\| \phi - \phi^{(n)} \right\| \to 0, \qquad n \to \infty.$$

a) Assume $|\phi| \leqslant M$ and ϕ is continuous in t. For each n, let

$$\phi_t^{(n)} = \begin{cases} n \displaystyle\int_{(j-1)/n}^{j/n} \phi_t dt, & \text{when } \dfrac{j}{n} < t \leqslant \dfrac{j+1}{n} \text{ and } 1 \leqslant j \leqslant n^2, \\ 0, & \text{else.} \end{cases}$$

By Cauchy-Schwarz inequality,

$$\int_{j/n}^{(j+1)/n} \left| \phi_t^{(n)} \right|^2 dt = n \left| \int_{(j-1)/n}^{j/n} \phi_t dt \right|^2 \leqslant \int_{(j-1)/n}^{j/n} \phi_t^2 dt,$$

so that

$$\int_T^\infty \left|\phi_t^{(n)}\right|^2 dt \leqslant \int_{T-2/n}^\infty \phi_t^2 dt. \tag{7.4}$$

Thus

$$\int_0^\infty \left|\phi_t - \phi_t^{(n)}\right|^2 dt = \int_0^T \left|\phi_t - \phi_t^{(n)}\right|^2 dt + \int_T^\infty \left|\phi_t - \phi_t^{(n)}\right|^2 dt$$

$$\leqslant \int_0^T \left|\phi_t - \phi_t^{(n)}\right|^2 dt + 4 \int_{T-2/n}^\infty \phi_t^2 dt. \tag{7.5}$$

Note that the continuity of ϕ implies that ϕ is bounded and uniformly continuous on $[0, T]$. Therefore by letting $n \to \infty$, the first term on the right-hand side of (7.5) converges to 0. Next, note that $\int_0^\infty \phi_t^2 dt < \infty$, a.s. Therefore by letting $T \to \infty$, we get that

$$\int_0^\infty \left|\phi_t - \phi_t^{(n)}\right|^2 dt \to 0, \quad \text{a.s.} \quad \text{as} \quad n \to \infty.$$

By (7.4), we have

$$\int_0^\infty \left|\phi_t - \phi_t^{(n)}\right|^2 dt \leqslant 4 \int_0^\infty |\phi_t|^2 dt < \infty.$$

Thus the dominated convergence theorem implies $\left\| \phi - \phi^{(n)} \right\| \to 0$.

b) We now remove the continuity assumption of ϕ. Assume $|\phi| \leqslant M$. For every n, let ψ_n be a nonnegative continuous function on $\mathbb{R}$ such that: (i) $\psi_n(t) = 0$ if $t \leqslant -1/n$ or $t \geqslant 0$, (ii) $\int_{-\infty}^\infty \psi_n(t) dt = 1$. Set

$$\phi_t^{(n)} = \int_0^t \psi_n(s - t)\phi_s ds.$$

Then $\phi^{(n)}$ is continuous, and $\left|\phi_t^{(n)}\right| \leqslant M$. It is easy to see $\phi_t^{(n)} \in \mathscr{F}_t$, and

$$\int_0^\infty \left|\phi_t - \phi_t^{(n)}\right|^2 dt \to 0, \quad n \to \infty.$$

Therefore the dominated convergence theorem implies that $\left\| \phi - \phi^{(n)} \right\| \to 0$. $\square$

Properties of Stochastic Integral

Proposition 7.4. Define

$$X_t = \int_0^t \phi_s dB_s := \int_0^\infty \phi_s dB_{s \wedge t}$$

for $\phi \in \mathscr{L}^2$. It has the following properties.

(1) X_t is continuous in t.

(2) $(X_t, \mathscr{F}_t)$ is a martingale and $\mathbb{E}X_t = 0$.

(3) Let

$$Y_t = \int_0^t \psi_s \mathrm{d}B_s.$$

Then

$$\mathbb{E}X_t Y_t = \mathbb{E}\int_0^t \phi_s \psi_s \mathrm{d}s,$$

thus $\|\phi\|^2 = \mathbb{E}I_\phi^2$ for every $\phi \in \mathscr{L}^2$.

(4) We have

$$\int_s^t (a\phi_u + b\psi_u)\mathrm{d}B_u = a\int_s^t \phi_u \mathrm{d}B_u + b\int_s^t \psi_u \mathrm{d}B_u$$

for any random variables $a, b \in \mathscr{F}_s$.

7.2 Itô's Formula

As the chain rule in calculus, the corresponding rule in the stochastic integral is the so-called Itô formula.

Assume u is $\mathscr{F}_t$-adapted and $v \in \mathscr{L}^2$. Let

$$X_t = X_0 + \int_0^t u_s \mathrm{d}s + \int_0^t v_s \mathrm{d}B_s, \qquad (7.6)$$

or formally write it as

$$\mathrm{d}X_t = u_t \mathrm{d}t + v_t \mathrm{d}B_t. \qquad (7.7)$$

Theorem 7.5. Assume that X_t satisfies (7.6) (or (7.7)) and function $f : \mathbb{R}_+ \times \mathbb{R} \to \mathbb{R}$ is twice differentially continuous. Let $Y_t = f(t, X_t)$. Then

$$\mathrm{d}Y_t = \frac{\partial f}{\partial t}(t, X_t)\mathrm{d}t + \frac{\partial f}{\partial x}(t, X_t)\mathrm{d}X_t + \frac{1}{2}\frac{\partial^2 f}{\partial x^2}(t, X_t)(\mathrm{d}X_t)^2,$$

where the squares and products for $\mathrm{d}t$ and $\mathrm{d}B_t$ are given by

product	$\mathrm{d}t$	$\mathrm{d}B_t$
$\mathrm{d}t$	0	0
$\mathrm{d}B_t$	0	t

Namely,

$$\mathrm{d}Y_t = \left[\frac{\partial f}{\partial t}(t, X_t) + u_t\frac{\partial f}{\partial x}(t, X_t) + \frac{1}{2}v_t^2\frac{\partial^2 f}{\partial x^2}(t, X_t)\right]\mathrm{d}t + v_t\frac{\partial f}{\partial x}(t, X_t)\mathrm{d}B_t.$$

Proof Without loss of generality, we assume

$$f, \ \frac{\partial f}{\partial t}, \ \frac{\partial f}{\partial x}, \ \frac{\partial^2 f}{\partial x^2}$$

are bounded. Take $0 = t_0 < t_1 < \cdots < t_n = t$. Taylor's expansion up to the second-order term gives

$$f(t, X_t) = f(0, X_0) + \sum_i \Delta f(t_i, X_i)$$

$$= f(0, X_0) + \sum_i \frac{\partial f}{\partial t} \Delta t_i + \sum_i \frac{\partial f}{\partial x} \Delta X_i + \frac{1}{2} \sum_i \frac{\partial^2 f}{\partial t^2} (\Delta t_i)^2$$

$$+ \sum_i \frac{\partial^2 f}{\partial t \partial x} (\Delta t_i)(\Delta X_i) + \frac{1}{2} \sum_i \frac{\partial^2 f}{\partial x^2} (\Delta X_i)^2 + \sum_i R_i,$$

where

$$\Delta g(t_i, X_i) = g(t_{i+1}, X_{t_{i+1}}) - g(t_i, X_{t_i}),$$
$$\Delta t_i = t_{i+1} - t_i, \ \Delta X_i = X_{t_{i+1}} - X_{t_i},$$
$$R_i = o\left(|\Delta t_i|^2 + |\Delta X_i|^2\right).$$

Letting $\max_i |\Delta t_i| \to 0$, we obtain

$$\sum_i \frac{\partial f}{\partial t} \Delta t_i = \sum_i \frac{\partial f}{\partial t}(t_i, X_{t_i}) \Delta t_i \xrightarrow{L^2} \int_0^t \frac{\partial f}{\partial t}(s, X_s) ds, \qquad (7.8)$$

$$\sum_i \frac{\partial f}{\partial x} \Delta X_i = \sum_i \frac{\partial f}{\partial x}(t_i, X_{t_i}) \Delta X_i \xrightarrow{L^2} \int_0^t \frac{\partial f}{\partial x}(s, X_s) dX_s. \qquad (7.9)$$

Meanwhile,

$$\sum_i \frac{\partial^2 f}{\partial x^2} (\Delta X_i)^2 = \sum_i \frac{\partial^2 f}{\partial x^2} u_i^2 (\Delta t_i)^2 + 2 \sum_i \frac{\partial^2 f}{\partial x^2} u_i v_i (\Delta t_i)(\Delta B_i)$$

$$+ \sum_i \frac{\partial^2 f}{\partial x^2} v_i^2 (\Delta B_i)^2. \qquad (7.10)$$

By letting $\max_i |\Delta t_i| \to 0$, the first two terms on the right-hand side go to zero, and the last term converges in L^2 to

$$\int_0^t \frac{\partial^2 f}{\partial x^2}(s, X_s) v_s^2 ds.$$

To see the last convergence, denote $a(t) = \frac{\partial^2 f}{\partial x^2}(t, X_t) v^2(t)$, $a_i = a(t_i)$. It suffices to show that

$$\mathbb{E} \left(\sum_i a_i (\Delta B_i)^2 - \sum_i a_i \Delta t_i \right)^2 = \sum_{i,j} \mathbb{E}\left[a_i a_j \left((\Delta B_i)^2 - \Delta t_i\right) \left((\Delta B_j)^2 - \Delta t_j\right)\right]$$

converges to zero. When $i > j$, $a_i a_j ((\Delta B_i)^2 - \Delta t_i)$ and $(\Delta B_j)^2 - \Delta t_j$ are independent, having mean zero. In the same way, when $i < j$, they have mean zero. Besides, when $\max_i |\Delta t_i| \to 0$,

$$\sum_i \mathbb{E} a_i^2 \left((\Delta B_i)^2 - \Delta t_i \right)^2 = 2 \sum_i \mathbb{E} a_i^2 (\Delta t_i)^2 \to 0.$$

Therefore, in the sense of $L^2(\mathbb{P})$,

$$\sum_i a_i (\Delta B_i)^2 \to \int_0^t a(s) \mathrm{d}s.$$

A same argument gives that when $\max_i |\Delta t_i| \to 0$, $\sum_i R_i \to 0$ in $L^2(\mathbb{P})$. □

Multidimensional Itô's formula   Assume $B_t = (B_t^1, \cdots, B_t^n)^T$ is an n-dimensional BM^n. Let

$$\mathrm{d}X_t^i = b_i(t, X_t)\mathrm{d}t + \sum_{j=1}^n \sigma_{ij}(t, X_t)\mathrm{d}B_t^j, \quad i = 1, 2, \cdots, n.$$

Here we only present the multidimensional Itô's formula, and the detailed study is delayed to next chapter.

Theorem 7.6. Assume $f = (f_1, \cdots, f_m) : [0, \infty) \times \mathbb{R}^n \to \mathbb{R}^m$ is twice differentially continuous. Let $Y_t = f(t, X_t)$. Then for $1 \leqslant i \leqslant m$

$$\mathrm{d}Y_t^i = \frac{\partial f_i}{\partial t}(t, X_t)\mathrm{d}t + \sum_j \frac{\partial f_i}{\partial x_j}(t, X_t)\mathrm{d}X_t^j + \sum_{j,k} \frac{1}{2}\frac{\partial^2 f_i}{\partial x_j \partial x_k}(t, X_t)\mathrm{d}X_t^j \mathrm{d}X_t^k,$$

where $\mathrm{d}B_t^j \mathrm{d}B_t^k = \delta_{jk}\mathrm{d}t$, $\mathrm{d}t\mathrm{d}t = 0$, $\mathrm{d}t\mathrm{d}B_t^j = 0$.

7.3   Stochastic Differential Equation (SDE) (Dimension One)

Stochastic differential equation (SDE) has the form

$$\mathrm{d}X_t = b(t, X_t)\mathrm{d}t + \sigma(t, X_t)\mathrm{d}B_t. \tag{7.11}$$

Its integral form is

$$X_t = X_0 + \int_0^t b(t, X_t)\mathrm{d}t + \int_0^t \sigma(t, X_t)\mathrm{d}B_t. \tag{7.12}$$

The solution X_t to this equation is called **Itô's process**, or **diffusion process**. In the following, we will study the existence and uniqueness for the solution of SDE.

Theorem 7.7. Assume that b and σ in (7.11) satisfy

$$|b(t,x)| + |\sigma(t,x)| \leqslant C(1+|x|), \qquad (7.13)$$

$$|b(t,x) - b(t,y)| + |\sigma(t,x) - \sigma(t,y)| \leqslant C|x-y|. \qquad (7.14)$$

Then the SDE (7.11) (or (7.12)) has a unique solution X_t, which is continuous in t.

Proof a) Prove the uniqueness first. Assume X_t and $\widetilde{X}_t$ are two solutions to SDE, with the initial condition $X_0 = Z$ and $\widetilde{X}_0 = \widetilde{Z}$, respectively. Let $\alpha_t = b(t,X_t) - b(t,\widetilde{X}_t), \beta_t = \sigma(t,X_t) - \sigma(t,\widetilde{X}_t)$. Then

$$\mathbb{E}\|X_t - \widetilde{X}_t\|^2 = \mathbb{E}\left[Z - \widetilde{Z} + \int_0^t \alpha_s \mathrm{d}s + \int_0^t \beta_s \mathrm{d}B_s\right]^2$$

$$\leqslant 3\,\mathbb{E}\left[Z - \widetilde{Z}\right]^2 + 3\,\mathbb{E}\left[\int_0^t \alpha_s \mathrm{d}s\right]^2 + 3\,\mathbb{E}\left[\int_0^t \beta_s \mathrm{d}B_s\right]^2$$

$$\leqslant 3\,\mathbb{E}\left[Z - \widetilde{Z}\right]^2 + 3\,t\,\mathbb{E}\left[\int_0^t \alpha_s^2 \mathrm{d}s\right] + 3\,\mathbb{E}\left[\int_0^t \beta_s^2 \mathrm{d}s\right]$$

$$\leqslant 3\,\mathbb{E}\left[Z - \widetilde{Z}\right]^2 + 3(1+t)C^2 \int_0^t \mathbb{E}|X_s - \widetilde{X}_s|^2 \mathrm{d}s.$$

Thus by Gronwall's lemma (cf. Exercise 15)

$$\mathbb{E}\|X_t - \widetilde{X}_t\|^2 \leqslant 3\,\mathbb{E}\left[Z - \widetilde{Z}\right]^2 e^{3(1+t)C^2 t}.$$

If $Z = \widetilde{Z}$, then $\mathbb{E}|X_t - \widetilde{X}_t|^2 = 0$ for every $t \geqslant 0$, so that

$$\mathbb{P}[X_t = \widetilde{X}_t \text{ for all rational } t \geqslant 0] = 1.$$

Therefore, by continuity,

$$\mathbb{P}[X_t = \widetilde{X}_t \text{ for every } t \geqslant 0] = 1.$$

b) Prove the existence, by using the iteration method. Let $Y_t^{(0)} = X_0$, and define inductively

$$Y_t^{(k+1)} = X_0 + \int_0^t b(s, Y_s^{(k)})\mathrm{d}s + \int_0^t \sigma\left(s, Y_s^{(k)}\right)\mathrm{d}B_s, \quad k \geqslant 0.$$

As proved above, we have

$$\mathbb{E}\left|Y_t^{(k+1)} - Y_t^{(k)}\right|^2 \leqslant 3(1+t)C^2 \int_0^t \mathbb{E}\left|Y_s^{(k)} - Y_s^{(k-1)}\right|^2 \mathrm{d}s, \quad k \geqslant 1.$$

Thus there exists constant A, depending only on C, t and $\mathbb{E}X_0^2$, such that

$$\mathbb{E}\left|Y_t^{(1)} - Y_t^{(0)}\right|^2 \leqslant At.$$

Thus by induction

$$\mathbb{E}\left|Y_t^{(k+1)} - Y_t^{(k)}\right|^2 \leqslant \frac{(At)^{k+1}}{(k+1)!}, \quad k \geqslant 0. \tag{7.15}$$

Note that

$$\sup_{0 \leqslant s \leqslant t}\left|Y_s^{(k+1)} - Y_s^{(k)}\right| \leqslant \sup_{0 \leqslant s \leqslant t}\int_0^s \left|b\left(r, Y_r^{(k)}\right) - b\left(r, Y_r^{(k-1)}\right)\right| dr$$
$$+ \sup_{0 \leqslant s \leqslant t}\left|\int_0^s \left[\sigma\left(r, Y_r^{(k)}\right) - \sigma\left(r, Y_r^{(k-1)}\right)\right] dB_r\right|.$$

Then by the martingale inequality (Corollary 5.9) and Lemma 7.1, we have

$$\mathbb{P}\left[\sup_{0 \leqslant s \leqslant t}\left|Y_s^{(k+1)} - Y_s^{(k)}\right| \geqslant 2^{-k}\right]$$
$$\leqslant \mathbb{P}\left[\sup_{0 \leqslant s \leqslant t}\int_0^s \left|b\left(r, Y_r^{(k)}\right) - b\left(r, Y_r^{(k-1)}\right)\right| dr \geqslant 2^{-k-1}\right]$$
$$+ \mathbb{P}\left[\sup_{0 \leqslant s \leqslant t}\left|\int_0^s \left[\sigma\left(r, Y_r^{(k)}\right) - \sigma\left(r, Y_r^{(k-1)}\right)\right] dB_r\right| \geqslant 2^{-k-1}\right]$$
$$\leqslant 4^{k+1}\mathbb{E}\left|\int_0^t b\left(r, Y_r^{(k)}\right) - b\left(r, Y_r^{(k-1)}\right) dr\right|^2$$
$$+ 4^{k+1}\mathbb{E}\left|\int_0^t \sigma\left(r, Y_r^{(k)}\right) - \sigma\left(r, Y_r^{(k-1)}\right) dB_r\right|^2$$
$$\leqslant 4^{k+1}t\mathbb{E}\int_0^t \left|b\left(r, Y_r^{(k)}\right) - b\left(r, Y_r^{(k-1)}\right)\right|^2 dr$$
$$+ 4^{k+1}\int_0^t \mathbb{E}\left|\sigma\left(r, Y_r^{(k)}\right) - \sigma\left(r, Y_r^{(k-1)}\right)\right|^2 dr.$$

And by conditions (7.13) and (7.14), we have

$$\mathbb{P}\left[\sup_{0 \leqslant s \leqslant t}\left|Y_s^{(k+1)} - Y_s^{(k)}\right| \geqslant 2^{-k}\right] \leqslant 4^{k+1}C^2(1+t)\int_0^t \mathbb{E}\left|Y_r^{(k)} - Y_r^{(k-1)}\right|^2 dr$$
$$\leqslant 4^{k+1}C^2(1+t)\int_0^t \frac{(At)^k}{k!} dr$$
$$\leqslant 4^{k+1}C^2(1+t)\frac{A^k t^{k+1}}{(k+1)!}.$$

Therefore

$$\sum_{k=1}^{\infty} \mathbb{P}\left[\sup_{0 \leqslant s \leqslant t}\left|Y_s^{(k+1)} - Y_s^{(k)}\right| \geqslant 2^{-k}\right] < \infty.$$

By Borel-Cantelli's lemma, we have

$$\mathbb{P}\left[\sup_{0 \leqslant s \leqslant t} \left|Y_s^{(k+1)} - Y_s^{(k)}\right| \geqslant 2^{-k}, \text{ i.o.}\right] = 0,$$

so for almost surely ω,

$$Y_t^{(n)}(\omega) = Y_t^{(0)}(\omega) + \sum_{k=0}^{n-1}\left(Y_t^{(k+1)}(\omega) - Y_t^{(k)}(\omega)\right)$$

converges to some X_t. Then by the continuity of $Y_t^{(n)}$ in t, we obtain the continuity of X_t. Furthermore, by the continuity of b and σ, we have

$$b\left(s, Y_s^{(n)}\right) \to b\left(s, X_s\right), \quad \sigma\left(s, Y_s^{(n)}\right) \to \sigma\left(s, X_s\right), \quad n \to \infty.$$

Thus

$$\int_0^t b\left(s, Y_s^{(n)}\right) \mathrm{d}s \to \int_0^t b\left(s, X_s\right) \mathrm{d}s.$$

From the dominated convergence theorem and Lemma 7.1, it follows that in the sense of $L^2(\mathbb{P})$ (and then in the almost sure sense)

$$\int_0^t \sigma\left(s, Y_s^{(n)}\right) \mathrm{d}s \to \int_0^t \sigma\left(s, X_s\right) \mathrm{d}s.$$

Therefore X_t satisfies SDE (7.12).   □

7.4   One-Dimensional Diffusion Process

In this section, we will use the SDE method to study one-dimensional diffusion processes.

Assume that $I = (\ell, r)$ is an open interval: $-\infty \leqslant \ell < r \leqslant \infty$. Suppose $\sigma(x)$ and $b(x)$ are C^1-functions on I, and $\sigma^2(x) > 0$ for each $x \in I$. The stochastic differential equation

$$\begin{cases} \mathrm{d}X_t = \sigma(X_t)\mathrm{d}B_t + b(X_t)\mathrm{d}t \\ X_0 = x \end{cases} \tag{7.16}$$

has a unique solution X_t^x until the explosion time $e = \lim_{n \uparrow \infty} \tau_n$, where

$$\tau_n = \inf\{t : X_t^x \notin [a_n, b_n]\}$$

(choose a_n, b_n such that $\ell < a_n < b_n < r$ such that $a_n \downarrow \ell$, $b_n \uparrow r$).

It is easy to prove that on set $[e < \infty]$, the limit $\lim_{t \uparrow e} X_t^x$ exists and equals ℓ or r a.e. Therefore, on set $\{e < \infty\}$, X_t^x is defined to be the limit for each $t \geqslant e$. Denote

$$L = a(x)\frac{\mathrm{d}^2}{\mathrm{d}x^2} + b(x)\frac{\mathrm{d}}{\mathrm{d}x}, \qquad a(x) = \frac{1}{2}\sigma^2(x), \tag{7.17}$$

then $(X_t^x)_{t \geqslant 0}$ is called **the minimal L-diffusion process**.

Uniqueness

The uniqueness means that the minimal L-diffusion process is the unique solution to the stochastic differential equation (7.16), that is $\mathbb{P}_x[e = \infty] = 1$ for every $x \in I$. We are going to introduce an explicit criterion for uniqueness.

Fix a reference point $\theta \in I$, and set

$$\kappa(x) = \int_\theta^x s(\xi) M(\xi) \mathrm{d}\xi, \qquad x \in I, \tag{7.18}$$

where

$$M(\xi) = \int_\theta^\xi \exp\left[\int_\theta^y \frac{b(z)}{a(z)} \mathrm{d}z\right] \frac{\mathrm{d}y}{a(y)}, \quad s(x) = \exp\left[-\int_\theta^x \frac{b(z)}{a(z)} \mathrm{d}z\right].$$

Next, set

$$S(x) = \int_\theta^x \exp\left[-\int_\theta^y \frac{b(z)}{a(z)} \mathrm{d}z\right] \mathrm{d}y. \tag{7.19}$$

Clearly, $S(x)$ is a strictly increasing function on I, satisfying

$$LS(x) = 0. \tag{7.20}$$

Indeed, from (7.20) we can easily obtain the expression (7.19) of S.

Lemma 7.8. Assume that $u(x)$ is the unique solution of

$$Lu(x) = u(x), \qquad u(\theta) = 1, \ u'(\theta) = 0. \tag{7.21}$$

Then

$$1 + \kappa(x) \leqslant u(x) \leqslant \exp(\kappa(x)), \qquad x \in I. \tag{7.22}$$

Proof See Exercise 18 of this chapter. □

Theorem 7.9.
   (1) If $\kappa(r-) = \kappa(\ell+) = \infty$, then

$$\mathbb{P}_x[e = \infty] = 1 \tag{7.23}$$

   for every $x \in I$.
   (2) If $\kappa(r-) < \infty$ or $\kappa(\ell+) < \infty$, then

$$\mathbb{P}_x[e < \infty] > 0 \tag{7.24}$$

   for all $x \in I$.

Proof (1) By using Itô's formula, we obtain that

$$de^{-t}u(X_t) = e^{-t}u'(X_t)\sigma(X_t)dB_t.$$

Hence $\{e^{-t}u(X_t)\}_{t\geqslant 0}$ is a martingale, thus the Doob's stopping theorem implies that $e^{-t\wedge\tau_n}u(X_{t\wedge\tau_n})$ is also a martingale. Let $n \to \infty$ to derive $Y_t := e^{-t\wedge e}u(X_{t\wedge e})$ is a nonnegative supermartingale. Therefore $\mathbb{E}Y_t \leqslant \mathbb{E}Y_0 = 1$.

Assume $\kappa(r-) = \kappa(\ell+) = \infty$. It follows from Lemma 7.8 that $u(r-) = u(\ell+) = \infty$, so $\mathbb{P}_x[e = \infty] = 1$. Otherwise there exists $K < \infty$ such that $\mathbb{P}_x[e \leqslant K] > 0$, which implies

$$\lim_{t\to\infty} \mathbb{E}Y_t \geqslant \liminf_{t\to\infty} \int_{[e\leqslant K]} e^{-t\wedge e}u(X_{t\wedge e})d\mathbb{P} = \infty.$$

This is impossible.

(2) Without loss of generality, assume that $\kappa(r-) < \infty$ and $\theta < x$. Let $\eta = \inf\{t : X_t = \theta\}$. Then by Lemma 7.8 we know that $e^{-t\wedge\eta\wedge e}u(X_{t\wedge\eta\wedge e})$ is a bounded martingale, so that

$$u(x) = \mathbb{E}_x\big[e^{-t\wedge\eta\wedge e}u(X_{t\wedge\eta\wedge e})\big].$$

Let $t \to \infty$ to get

$$u(x) = \mathbb{E}_x\Big[e^{-e}u(r-); \lim_{t\to e\wedge\eta} X_t = r\Big] + \mathbb{E}_x\Big[e^{-\eta}u(\theta); \lim_{t\to e\wedge\eta} X_t = \theta\Big].$$

Therefore it must hold

$$\mathbb{E}_x\Big[e^{-e}; \lim_{t\to e\wedge\eta} X_t = r\Big] > 0,$$

hence $\mathbb{P}_x[e < \infty] > 0$. Otherwise, if $\mathbb{E}_x[e^{-e}; \lim_{t\to e\wedge\eta} X_t = r] = 0$, then

$$u(x) = \mathbb{E}_x\Big[e^{-\eta}u(\theta); \lim_{t\to e\wedge\eta} X_t = \theta\Big] \leqslant u(\theta).$$

This is a contradiction, since u is strictly increasing. $\square$

Recurrence

Denote $T_y = \inf\{t \geqslant 0 : X_t = y\}$ for $y \in I$. The diffusion process X_t is said to be recurrent, if $\mathbb{P}_x[T_y < \infty] = 1$ for every $x, y \in I$.

Theorem 7.10.
(1) If $S(\ell+) = -\infty$ and $S(r-) = \infty$, then

$$\mathbb{P}_x[e = \infty] = \mathbb{P}_x\Big[\limsup_{t\uparrow\infty} X_t = r\Big] = \mathbb{P}_x\Big[\liminf_{t\uparrow\infty} X_t = \ell\Big] = 1 \quad (7.25)$$

for each $x \in I$.

(2) If $S(\ell+) > -\infty$ and $S(r-) = \infty$, then $\lim\limits_{t\uparrow e} X(t)$ exists $\mathbb{P}_x$-$a.s.$, and

$$\mathbb{P}_x\left[\lim_{t\uparrow e} X_t = \ell\right] = P_x\left[\sup_{t<e} X_t < r\right] = 1 \qquad (7.26)$$

for each $x \in I$.
By exchanging ℓ and r, we have the same conclusion.
(3) If $S(\ell+) > -\infty$ and $S(r-) < \infty$, then $\lim\limits_{t\uparrow e} X(t)$ $\mathbb{P}_x$-$a.s.$ exists, and for each $x \in I$, we have

$$\mathbb{P}_x\left[\lim_{t\uparrow e} X_t = \ell\right] = 1 - \mathbb{P}_x\left[\lim_{t\uparrow e} X_t = r\right] = \frac{S(r-) - S(x)}{S(r-) - S(\ell+)}. \qquad (7.27)$$

Furthermore, we have

Theorem 7.11. For every $x \in I$, $\mathbb{P}_x(e < \infty) = 1$ iff one of the following conditions holds;
(1) $\kappa(r-) < \infty$ and $\kappa(\ell+) < \infty$;
(2) $\kappa(r-) < \infty$ and $S(\ell+) = -\infty$;
(3) $\kappa(\ell+) < \infty$ and $S(r-) = \infty$.

The proofs of Theorems 7.10-7.11 can be found in [39; §6.3].

Ergodicity

A probability measure π on I is called the **stationary distribution** of X_t, if the initial distribution of X_0 is π, then it holds that $X_t \sim \pi$ for every $t \geqslant 0$, that is for every Borel set B in I,

$$\pi(B) = \int_I P_t(x, B)\pi(\mathrm{d}x),$$

where $P_t(x, B) = \mathbb{P}_x[X_t \in B]$.
Suppose $\pi(\mathrm{d}x) = \pi(x)\mathrm{d}x$, that is the probability measure π has density function $\pi(x)$ with respect to Lebesgue measure. Then

$$\pi(y) = \int_I p_t(x, y)\pi(x)\mathrm{d}x,$$

where $p_t(x, y)$ is the density function $P_t(x, \mathrm{d}y)$ with respect to Lebesgue measure.
Indeed, one-dimensional diffusion process is symmetric, that is there exists measure μ such that

$$\int_I fLg\mathrm{d}\mu = \int_I gLf\mathrm{d}\mu, \quad f, g \in C^2(I). \qquad (7.28)$$

Therefore once $\mu(I) < \infty$, then we obtain the stationary distribution.

It is easy to check that

$$\mu(A) = \int_A \frac{dy}{a(y)} \exp\left[\int_\theta^x \frac{b(z)}{a(z)} dz\right]$$

satisfies (7.28), where $c \in I$ is arbitrary. Thus if

$$Z = \int_I \frac{dy}{a(y)} \exp\left[\int_\theta^x \frac{b(z)}{a(z)} dz\right] < \infty, \tag{7.29}$$

then

$$\pi(y) = \frac{1}{Za(y)} \exp\left[\int_\theta^x \frac{b(z)}{a(z)} dz\right]$$

which is the density of stationary distribution π. Therefore the diffusion process X_t is ergodic if and only if

$$\kappa(\ell+) = \kappa(r-) = \infty \quad \text{and} \quad \int_I \frac{dy}{a(y)} \exp\left[\int_\theta^x \frac{b(z)}{a(z)} dz\right] < \infty.$$

Estimate of First Eigenvalue

Let $L = a(x)\dfrac{d^2}{dx^2} + b(x)\dfrac{d}{dx}$ and consider the L-diffusion process on the half line $[0, D)\,(D \leqslant \infty)$. We are going to study the estimate for the first eigenvalue with Neumann boundary. Assume $a(x) > 0$ for $x \in [0, D)$ and $Z < \infty$, where Z is defined by (7.29) and $I = (0, D)$. That is the diffusion process X_t is ergodic. Denote $C(x) = \int_0^x b/a$, and set

$$\pi(dx) = \frac{1}{Za(x)} \exp[C(x)]dx.$$

On $L^2(\pi)$, the operator L has a trivial eigenvalue $\lambda_0 = 0$ with constant eigenfunction 1. What we are interested in is the next eigenvalue λ_1, that is, the so-called (nontrivial) first eigenvalue. For the case of finite interval $(D < \infty)$ and continuous coefficients, λ_1 is just the minimal λ such that the eigen-equation $Lf = -\lambda f$ holds for non-constant function f. In general, λ_1 is defined by the classical variational formula (the min-max principle):

$$\lambda_1 = \inf\left\{\int_0^D a(x)f'(x)^2\pi(dx) : f \in C^1[0, D), \quad \pi(f) = 0, \quad \pi(f^2) = 1\right\},$$

where $\pi(f) = \int_0^D f d\pi$.

Theorem 7.12. Let $\mathscr{F} = \{f : f'(x) > 0,\ x \in (0, D),\ \pi(f) \geqslant 0\}$. Then

$$\lambda_1 \geqslant \sup_{f \in \mathscr{F}} \inf_{x \in (0,D)} \left[\frac{e^{-C(x)}}{f'(x)} \int_x^D \frac{f(u)e^{C(u)}}{a(u)} du \right]^{-1}. \tag{7.30}$$

Furthermore, if the coefficients a and b are continuous, then the equality in (7.30) holds.

Proof For $f \in \mathscr{F}$, denote

$$I(f)(x) = \frac{e^{-C(x)}}{f'(x)} \int_x^D \frac{f(u)e^{C(u)}}{a(u)} du.$$

Let $g \in C^1[0, D)$ satisfy $\pi(g) = 0$ and $\pi(g^2) = 1$. Then for every $f \in \mathscr{F}$, we have

$$1 = \frac{1}{2} \int_0^D \pi(dx)\pi(dy)[g(y) - g(x)]^2$$

$$= \int_{[x \leqslant y]} \pi(dx)\pi(dy) \left(\int_x^y g'(u)\sqrt{f'(u)}/\sqrt{f'(u)}\, du \right)^2$$

$$\leqslant \int_{[x \leqslant y]} \pi(dx)\pi(dy) \int_x^y g'(u)^2 f'(u)^{-1} du \int_x^y f'(u)du$$

(by Cauchy-Schwarz inequality)

$$= \int_{[x \leqslant y]} \pi(dx)\pi(dy) \int_x^y a(u)g'(u)^2 e^{C(u)} \frac{e^{-C(u)}}{a(u)f'(u)} du[f(y) - f(x)]$$

$$= \int_0^D a(u)g'(u)^2 \pi(du) \frac{Ze^{-C(u)}}{f'(u)} \int_0^u \pi(dx) \int_u^D \pi(dy)[f(y) - f(x)]. \tag{7.31}$$

But

$$\int_0^u \pi(dx) \int_u^D \pi(dy)[f(y) - f(x)]$$

$$= \int_0^u \pi(dx) \int_u^D f(y)\pi(dy) - \int_0^u f(x)\pi(dx) \int_u^D \pi(dy)$$

$$= \int_u^D f(y)\pi(dy) - \int_u^D \pi(dx) \int_u^D f(y)\pi(dy) - \int_0^u f(x)\pi(dx) \int_u^D \pi(dy)$$

$$= \int_u^D f(y)\pi(dy) - \int_u^D \pi(dx) \int_0^D f(y)\pi(dy) \qquad \text{(since } \pi(f) \geqslant 0)$$

$$\leqslant \int_u^D f(y)\pi(dy) = \frac{1}{Z} \int_u^D \frac{f(y)e^{C(y)}}{a(y)} dy.$$

Combine this equation with (7.31) to derive

$$\int_0^D a(x)g'(x)^2\pi(\mathrm{d}x) \geqslant \inf_{x\in(0,D)} I(f)(x)^{-1}.$$

First make infimum in $g \in C^1[0, D)$ with $\pi(g) = 0$ and $\pi(g^2) = 1$, and then make the supremum in $f \in \mathscr{F}$ to derive (7.30).

To prove the equality in (7.30), we need more information about the eigenfunctions, for this refer to [16; Proposition 6.4, Lemma 6.2], [8; Proof of Theorem 1.1] and [9].   □

In this section and §3.2, we have studied a typical stability speed, i.e. the principal eigenvalue of the corresponding generator. Fortunately, a rather complete picture has become clearer recently for the speed of various types of stability in dimension one. Refer to the survey paper [14] for details.

Finally, in view of the study here on the stability speed for stochastic processes, as well as the economic/Google's search studied in §1.1, the importance of computing the first nontrivial eigenpair (eigenvalue and its corresponding eigenvector) should be clear. Refer to [13] for more information.

7.5   Supplements and Exercises

(1) Compute directly: (a) $\int_0^t B_s\mathrm{d}B_s$, (b) $\int_0^t s\mathrm{d}B_s$.

(2) Assume $\phi_s(\omega) \equiv \phi_s$ for every $s \geqslant 0$ and f has bounded variation in $[0,t]$. Prove

$$\int_0^t \phi_s\mathrm{d}B_s = tB_t - \int_0^t B_s\mathrm{d}\phi_s.$$

(3) Assume $0 = t_0 < t_1 < \cdots < t_{n+1} = 1, \theta_i \in (t_i, t_{i+1}]$. Let $\xi_t^{(n)} = B_{\theta_i}$ for $t \in (t_i, t_{i+1}]$. Try to prove that when the diameter of division $\{t_0, t_1, \cdots, t_{n+1}\}$ tends to zero with probability 1, it holds that

$$\sup_{0\leqslant t\leqslant 1} \left|\xi_t^{(n)} - B_t\right| \to 0, \qquad n \to \infty.$$

But when $\{\theta_i\}_{i\geqslant 1}$ varies, the limit points of summation

$$\mathbb{E} \sum_{k=0}^n B_{\theta_k}\left(B_{t_{k+1}} - B_{t_k}\right) = \sum_{k=0}^n (\theta_k - t_k)$$

can fill the whole $[0, 1]$.

(4) Replacing the original sum by the symmetric sum

$$\sum_{i=0}^{n-1} \frac{\xi_i + \xi_{i+1}}{2} \left(B_{t_{i+1}} - B_{t_i} \right)$$

approximating the stochastic integral, we obtain the **Stratonovich integral**, and denote by $\int_0^\infty \phi_t \circ dB_t$. Its differential form is $\phi_t \circ dB_t$. Prove that the Itô's formula for the Stratonovich integral is of the usual form. Namely, assume that $f : \mathbb{R}_+ \times \mathbb{R} \to \mathbb{R}$ is differential in t and third continuously differential in x. Let $Y_t = f(t, X_t)$. Then

$$dY_t = \frac{\partial f}{\partial t} dt + \frac{\partial f}{\partial x} \circ dX_t.$$

The advantage of this integral is its symmetry, while its disadvantage is the stricter conditions on f. However its concrete calculation requires Itô's integral.

(5) Prove the properties in Proposition 7.4.

Hint: By Lemma 7.1 and Proposition 7.2, and Exercise 19 in Chapter 5, we need prove the assertions for "step-like" functions.

(6) Assume $B_t = (B_t^1, \cdots, B_t^d)$ is d-dimensional BM^d. Let $R_t = |B_t|$ be the radial process. Then

$$R_t dR_t = \sum_{i=1}^{d} B_t^i dB_t^i + \frac{n-1}{2} dt.$$

(7) Consider the following Langevin equation:

$$dX_t = \sigma dB_t - bX_t dt,$$

where σ and b are constants.

(a) Use Itô's formula to solve the equation.
   Hint: Consider function $f(t, x) = e^{bt} x$.
(b) Given $X_0 = x$, find the distribution $P(t, x, \cdot)$ of X_t.
(c) Let $\sigma^2 = 1$ and $b = 1/2$. Prove $\lim_{t \to \infty} P(t, x, \cdot)$ is the standard normal distribution.

(8) Given constants α and β, prove the following population equation

$$dN_t = \alpha N_t dB_t + \beta N_t dt, \quad t \geqslant 0$$

has solution $N_t = N_0 e^{bt + \alpha B_t}$, where $b = \beta - \alpha^2/2$.
Hint: Consider function $f(t, x) = e^{bt + \alpha x}$.

(9) Consider the solutions of the following two non-stochastic equation.

(a) Prove the unique solution to $dX_t = X_t^2 dt$ with $X_0 = 1$ is

$$X_t = \frac{1}{1-t}, \quad 0 \leqslant t < 1.$$

Thus prove that $X_t \to \infty$ (explosive) when $t \uparrow 1$.

(b) Prove the solutions to $dX_t = 3X_t^{2/3} dt$ with $X_0 = 0$ are not unique. Indeed, the solutions are as follows:

$$X_t = \begin{cases} (t-a)^3, & t > a; \\ 0, & t \leqslant a, \end{cases}$$

where $a > 0$.

(10) Prove the diffusion process X_t of $dX_t = cX_t^r dt$ is not explosive iff $0 < r < 1$, where c is a constant.

(11) Assume $\phi \in \mathscr{L}^2$ and let $X_t = \int_0^t \phi_s dB_s$. Then

$$\mathbb{P}\left[\sup_{0 \leqslant t \leqslant T} |X_t| > \lambda\right] \leqslant \frac{1}{\lambda^2} \mathbb{E}\int_0^T \phi_s^2 ds$$

for every λ and $T > 0$.

(12) Let $dX_t = \mu_1(t, X_t)dt + \sigma_1(t, X_t)dB_t$ and $dY_t = \mu_2(t, Y_t)dt + \sigma_2(t, Y_t)dB_t$. Prove

(a) $dX_t^2 = \sigma_1(t, X_t)^2 dt + 2X_t dX_t$;

(b) $d(X_t Y_t) = X_t dY_t + Y_t dX_t + \sigma_1(t, X_t)\sigma_2(t, Y_t)dt$.

(13) Let $Y_t = \exp(\mu t + \sigma B_t), \mu, \sigma \in \mathbb{R}$. Find dY_t.

(14) Let $Y_t = e^{iB_t}$. Then $Y_t = X_t^{(1)} + iX_t^{(2)}$ satisfies

$$dX_t^{(1)} = -\frac{1}{2}X_t^{(1)}dt - X_t^{(2)}dB_t, \quad dX_t^{(2)} = -\frac{1}{2}X_t^{(2)}dt + X_t^{(1)}dB_t.$$

(15) (Gronwall's lemma) Assume ϕ and f are nonnegative Borel functions on $[0, \infty)$, such that

$$\phi(t) \leqslant c + \int_0^t f(s)\phi(s)ds, \quad t \geqslant 0,$$

where $c \geqslant 0$ is constant. Then

$$\phi(t) \leqslant c \exp\left[\int_0^t f(s)ds\right], \quad t \geqslant 0.$$

(A recent extension of this lemma refers to [51], see also Appendix A in [11].)

(16) Prove the solution to the following one-dimensional diffusion process on $\mathbb{R}$ is not explosive:

$$\mathrm{d}X_t = \sigma(X_t)\mathrm{d}B_t,$$

where $\sigma > 0$.

(17) Let L be the differential operator give by (7.17). Prove that for every $f \in C^2((\ell, r))$

$$Lf(x) = \frac{\mathrm{d}}{\mathrm{d}M(x)} \frac{\mathrm{d}}{\mathrm{d}S(x)} f(x), \quad x \in (\ell, r).$$

(18) Prove Lemma 7.8.

Hint: First prove that (7.21) is equivalent to

$$u(x) = 1 + \int_c^x \mathrm{d}S(x) \int_c^y u(z)\mathrm{d}M(z),$$

and then use the iteration

$$u(x) \equiv 0, u_n(x) = \int_c^x \mathrm{d}S(x) \int_c^y u_{n-1}(z)\mathrm{d}M(z),$$

to prove $u(x) = \sum_{n=0}^{\infty} u_n(x)$.

(19) Prove Theorem 7.10.

Hint: Consider $\mathrm{d}S(X_t)$.

Chapter 8

Semimartingale and Stochastic Integral

8.1   Uniqueness of Doob-Meyer Decomposition

In this section, we are preparing for establishing more general stochastic integrals.

Assume $(\Omega, \mathscr{F}, \mathbb{P})$ is probability space and $(\mathscr{F}_t)_{t \geqslant 0}$ is an increasing σ-filtration. The Doob-Meyer decomposition says that every submartingale $(X_t, \mathscr{F}_t, \mathbb{P})$ can be expressed as:

$$X_t = M_t + A_t,$$

where $(M_t, \mathscr{F}_t, \mathbb{P})$ is a martingale and (A_t) is an increasing process:

$$s \leqslant t \Longrightarrow A_s \leqslant A_t, \qquad \text{a.s.}$$

However we only deal with a special case. Assume $(X_t, \mathscr{F}_t, \mathbb{P})$ is a martingale and right-continuous and progressively measurable. Notice for the right-continuous process, the adaption implies the progressive measurability. Then $(X_t^2, \mathscr{F}_t, \mathbb{P})$ is a submartingale. We will prove the following decomposition

$$X_t^2 = M_t + A_t.$$

For the special case $X_t = B_t$, since $B_t^2 - t$ is martingale, we already have the decomposition: $b_t^2 = M_t + A_t$, where

$$M_t = B_t^2 - t, \qquad A_t = t.$$

We study the uniqueness first, and then the existence. Hereafter, we always assume the martingales and submartingales are right-continuous and progressively measurable, and for locally bounded variation function f, we denote by $|f|(T)$ the total variation of f on $[0, T]$.

The following assertion is a kind of integration by parts formula (see Exercise 1 in this chapter).

Lemma 8.1. Assume $(X_t, \mathscr{F}_t, \mathbb{P})$ is a martingale and $A : [0, \infty) \times \Omega \to \mathbb{R}$ is progressively measurable and a.s. continuous with locally bounded variation. Suppose

$$\mathbb{E}\left[\left(\sup_{0 \leqslant t \leqslant T} |X_t|\right)(|A|(T) + |A_0|)\right] < \infty, \quad T > 0.$$

Then $\left(X_t A_t - \int_0^t X_s A(\mathrm{d}s), \mathscr{F}_t, \mathbb{P}\right)$ is martingale.

Proof Assume $0 \leqslant s < t$ and $\Gamma \in \mathscr{F}_s$. Denote $u_{n,k} = s + k(t - s)/n$ for $0 \leqslant k \leqslant n$. Then

$$\mathbb{E}[X_t A_t - X_s A_s; \Gamma]$$

$$= \mathbb{E}\left[\sum_{k=0}^{n-1} \left(X_{u_{n,k+1}} A_{u_{n,k+1}} - X_{u_{n,k}} A_{u_{n,k}}\right); \Gamma\right]$$

$$= \mathbb{E}\left[\sum_{k=0}^{n-1} X_{u_{n,k+1}} \left[A_{u_{n,k+1}} - A_{u_{n,k}}\right]; \Gamma\right]. \quad \text{(by the martingale property)}$$

Since $u \to X_u$ is right-continuous, $u \to A_u$ is a.s. continuous with locally bounded variation, thus

$$\sum_{k=0}^{n-1} \left[X_{u_{n,k+1}}\left(A_{u_{n,k+1}} - A_{u_{n,k}}\right)\right] \to \int_s^t X_s A(\mathrm{d}s), \quad \mathbb{P}\text{-a.s.}$$

By the assumption, we see that this convergence is really in $L^1(\mathbb{P})$.   □

Lemma 8.2. Assume $(X_t, \mathscr{F}_t, \mathbb{P})$ is an a.s. continuous martingale. Define $\zeta = \sup\{t \geqslant 0 : |X|(t) < \infty\}$. Then with probability 1, we have $X_{t \wedge \zeta} = X_0$, $t \geqslant 0$ for all $0 \leqslant s < t$. In particular, if $\mathbb{P}[X_t = X_s] = 0$, then with probability 1, $t \to X_t$ has unbounded variation in every interval.

Proof Without loss of generality, assume $X_0 \equiv 0$ and X_t is continuous everywhere. Define

$$\zeta_R = \sup\{t \geqslant 0 : |X|(t) < R\},$$
$$= \inf\{t \geqslant 0 : |X|(t) \geqslant R\}, \quad R > 0.$$

(Notice $|X|(t)$ is increasing in t.) Then ζ_R is a stopping time and $\zeta_R \uparrow \zeta$ as $R \uparrow \infty$. From Lemma 8.1 and the Doob's stopping theorem, it follows that

$$\left(X_{t \wedge \zeta_R}^2 - \int_0^{t \wedge \zeta_R} X_s X(\mathrm{d}s), \mathscr{F}_t, \mathbb{P}\right)$$

is a martingale. Therefore

$$\mathbb{E} X_{t \wedge \zeta_R}^2 = \mathbb{E}\int_0^{t \wedge \zeta_R} X_s X(\mathrm{d}s) \quad (\text{since} \quad X_0 \equiv 0).$$

On the other hand, since $X_{\bullet \wedge \zeta_R}$ is a.s. continuous with locally bounded variation,

$$X^2_{t \wedge \zeta_R} = 2 \int_0^{t \wedge \zeta_R} X_s X(\mathrm{d}s), \qquad \mathbb{P}\text{-a.s.}$$

The above two equations imply

$$\mathbb{E}X^2_{t \wedge \zeta_R} = 0, \qquad t \geqslant 0$$
$$\Longrightarrow X_{t \wedge \zeta_R} = 0, \qquad t \geqslant 0, \qquad \mathbb{P}\text{-a.s.}$$
$$\Longrightarrow X_{t \wedge \zeta} = 0 = X_0, \qquad t \geqslant 0, \qquad \mathbb{P}\text{-a.s.}$$

To prove the last assertion, let

$$^s X_\bullet = X_\bullet - X_{\bullet \wedge s}, \qquad s \geqslant 0.$$

We need only prove: for every $t > s$,

$$|^s X|(t) = \infty, \qquad \mathbb{P}\text{-a.s.}$$

For this, set

$$^s \zeta = \sup\{u \geqslant 0 : |^s X|(u) < \infty\}.$$

Then

$$\mathbb{P}[\,|^s X|(t) < \infty\,] = \mathbb{P}[^s\zeta > t].$$

Apply the first assertion of the lemma to the process $^s X$ to derive

$$^s X_{t \wedge \,^s\zeta} = {}^s X_0 = 0, \qquad \text{a.s.}$$
$$\Longrightarrow {}^s X_t = 0 \quad \text{on } [^s\zeta > t], \qquad \text{a.s.}$$
$$\Longrightarrow X_t = X_s \quad \text{on } [^s\zeta > t], \qquad \text{a.s.}$$
$$\Longrightarrow \mathbb{P}[^s\zeta > t] \leqslant \mathbb{P}[X_t = X_s] = 0. \qquad \square$$

Corollary 8.3. Given $(X_t)_{t \geqslant 0}$, possibly except a null measure set, there exists at most one right-continuous and progressively measurable process (A_t) such that
(1) $A_0 \equiv 0$, (A_t) is a.s. continuous with locally bounded variation,
(2) $(X_t^2 - A_t, \mathscr{F}_t, \mathbb{P})$ is a martingale.

Proof Let two processes (A_t) and (A'_t) satisfy the above condition. Then $(X_t - A_t) - (X_t - A'_t) = A_t - A'_t$ is a martingale. By applying Lemma 8.2, we have the process $(A_t - A'_t)_{t \geqslant 0}$ to derive

$$A_t - A'_t = A_0 - A'_0 = 0, \qquad \mathbb{P}\text{-a.s.} \quad \square$$

8.2   Existence of Doob-Meyer Decomposition

Theorem 8.4 (Doob-Meyer). Assume $(X_t, \mathscr{F}_t, \mathbb{P})$ is a.s. continuous $L^2(\mathbb{P})$-martingale. Then there exists $\mathbb{P}$-a.s. unique right-continuous and progressively measurable process (A_t), such that

(1) $A_0 = 0$, A_t is non-decreasing and a.s. continuous,

(2) $(X_t^2 - A_t, \mathscr{F}_t, \mathbb{P})$ is a martingale.

Proof The uniqueness is Corollary 8.3. We now prove the existence. Without loss of generality, assume $X_0 \equiv 0$. Define successively a sequence of doubly indexed stopping times $(\tau_k^n)_{k \geqslant 0}^{n \geqslant 0}$ as follows. At step 0, let

$$\tau_k^0 \equiv k, \qquad k \geqslant 0; \qquad (\Longrightarrow \tau_k^0 < \tau_{k+1}^0).$$

At step $n \geqslant 1$, we start $\tau_\bullet^n$ at $\tau_0^n \equiv 0$. Next, define the other $\{\tau_\ell^n\}_{\ell \geqslant 1}$ inductively. Suppose that we have already defined $\tau_\ell^n (\ell \geqslant 0)$ and let

$$\tau_k^{n-1} \leqslant \tau_\ell^n < \tau_{k+1}^{n-1}$$

for some $k \geqslant 0$ (unique). Then define

$$\tau_{\ell+1}^n = \left(\inf \left\{ t \geqslant \tau_\ell^n : \sup_{\tau_\ell^n \leqslant s \leqslant t} |X_s - X_{\tau_\ell^n}| \geqslant 1/n \right\} \right) \wedge (\tau_\ell^n + 1/n) \wedge \tau_{k+1}^{n-1}.$$

Inductively on n we get $\tau_k^n < \tau_{k+1}^n$ for $n \geqslant 0$, $k \geqslant 0$, and $\{\tau_k^n\} \subset \{\tau_k^{n+1}\}$, $\mathbb{P}$-a.s. Furthermore, it follows from the a.s. continuity that

$$\text{when } k \to \infty, \ \tau_k^n \to \infty, \qquad \mathbb{P}\text{-a.s.}$$

$$\tau_k^n \leqslant t \leqslant \tau_{k+1}^n \Longrightarrow |X_t - X_{\tau_k^n}| \leqslant 1/n, \qquad k \geqslant 0, \quad \mathbb{P}\text{-a.s.}$$

Take a sequence $K_n \uparrow \infty$ such that $\mathbb{P}[\tau_{K_n}^n \leqslant n] \leqslant 1/n$, and define

$$M_t^n = \sum_{k=0}^{K_n} X_{\tau_k^n}\left(X_{t \wedge \tau_{k+1}^n} - X_{t \wedge \tau_k^n} \right), \qquad A_t^n = \sum_{k=0}^{K_n} \left(X_{t \wedge \tau_{k+1}^n} - X_{t \wedge \tau_k^n} \right)^2.$$

It is easy to check the following facts:

(a) $M_0^n = A_0^n = 0$, $n \geqslant 0$;

(b) $(M_t^n, \mathscr{F}_t, \mathbb{P})$ is an a.s. continuous martingale;

(c) $A_t^n \geqslant 0$ is progressively measurable and a.s. continuous; if $t \geqslant s + 1/n$, then $A_s^n \leqslant A_t^n$; (Notice if s, t is in the same $[\tau_k^n, \tau_{k+1}^n)$, then the increasing property may fail.)

(d)

$$X_t(\omega)^2 = 2M_t^n(\omega) + A_t^n(\omega)$$

for every $n \geqslant 1$, $0 \leqslant t \leqslant n$, $\omega \in \Lambda_n := [\tau_{K_n}^n > n]$.

Next we prove

$$\lim_{m\to\infty} \sup_{n\geqslant m} \mathbb{P}\Big[\sup_{0\leqslant t\leqslant T} \big|M_t^n - M_t^m\big| \geqslant \varepsilon\Big] = 0 \tag{8.1}$$

for every $T > 0$ and $\varepsilon > 0$. For this, assume $n > m \geqslant T$, and set $\zeta = T \wedge \tau_{K_m}^m \wedge \tau_{K_n}^n$. Then

$$\mathbb{P}\Big[\sup_{0\leqslant t\leqslant T} \big|M_t^n - M_t^m\big| \geqslant \varepsilon\Big]$$

$$\leqslant \mathbb{P}\big[\tau_{K_m}^m \leqslant m \text{ or } \tau_{K_n}^n \leqslant n\big] + \mathbb{P}\Big[\sup_{0\leqslant t\leqslant T} \big|M_{t\wedge\zeta}^n - M_{t\wedge\zeta}^m\big| \geqslant \varepsilon\Big]$$

$$\leqslant 2/m + \mathbb{E}\big[(M_\zeta^n - M_\zeta^m)^2\big]/\varepsilon^2.$$

In the last step we have used Doob's inequality (Theorem 5.10). Define

$$\rho_k = \tau_k^m \wedge \zeta \quad \text{and} \quad \sigma_\ell = \tau_\ell^n \wedge \zeta.$$

Since $m < n$, $\{\rho_k\} \subset \{\sigma_\ell\}$. By definition, and noting $\zeta \leqslant \tau_m^m \wedge \tau_n^n$ and $\{\rho_k\} \subset \{\sigma_\ell\}$, we have

$$M_\zeta^n - M_\zeta^m = \sum_{\ell=0}^{K_n} X_{\tau_\ell^n}\Big(X_{\zeta\wedge\tau_{\ell+1}^n} - X_{\zeta\wedge\tau_\ell^n}\Big) - \sum_{k=0}^{K_m} X_{\tau_k^m}\Big(X_{\zeta\wedge\tau_{k+1}^m} - X_{\zeta\wedge\tau_k^m}\Big)$$

$$= \sum_{\ell=0}^{\infty} X_{\sigma_\ell}\Big(X_{\sigma_{\ell+1}} - X_{\sigma_\ell}\Big) - \sum_{k=0}^{\infty} X_{\rho_k}\Big(X_{\rho_{k+1}} - X_{\rho_k}\Big)$$

$$= \sum_{k,\ell=0}^{\infty} I_{[\rho_k,\rho_{k+1})}(\sigma_\ell) X_{\sigma_\ell}\Big(X_{\sigma_{\ell+1}} - X_{\sigma_\ell}\Big) - \sum_{k,\ell=0}^{\infty} I_{[\rho_k,\rho_{k+1})}(\sigma_\ell) X_{\rho_k}\Big(X_{\sigma_{\ell+1}} - X_{\sigma_\ell}\Big)$$

$$= \sum_{k,\ell=0}^{\infty} I_{[\rho_k,\rho_{k+1})}(\sigma_\ell)\Big(X_{\sigma_\ell} - X_{\rho_k}\Big)\Big(X_{\sigma_{\ell+1}} - X_{\sigma_\ell}\Big), \qquad \mathbb{P}\text{-a.s.}$$

By the martingale property, the summands in the last double sum are orthogonal in $L^2(\mathbb{P})$. Thus

$$\mathbb{E}\big[(M_\zeta^n - M_\zeta^m)^2\big] = \mathbb{E}\Big[\sum_{k,\ell=0}^{\infty} I_{[\rho_k,\rho_{k+1})}(\sigma_\ell)\Big(X_{\sigma_\ell} - X_{\rho_k}\Big)^2\Big(X_{\sigma_{\ell+1}} - X_{\sigma_\ell}\Big)^2\Big]$$

$$\leqslant \frac{1}{m^2}\mathbb{E}\Big[\sum_{k,\ell=0}^{\infty} I_{[\rho_k,\rho_{k+1})}(\sigma_\ell)\Big(X_{\sigma_{\ell+1}} - X_{\sigma_\ell}\Big)^2\Big]$$

$$= \frac{1}{m^2}\mathbb{E}\Big[\sum_{\ell=0}^{\infty} \Big(X_{\sigma_{\ell+1}} - X_{\sigma_\ell}\Big)^2\Big]$$

$$= \frac{1}{m^2}\mathbb{E}(X_\zeta^2) \quad \text{(by orthogonality)}$$

$$\leqslant \frac{1}{m^2}\mathbb{E}(X_T^2) < \infty.$$

This proves (8.1).

Now, let us apply the following lemma to complete the proof of existence.

Lemma 8.5. Assume that (X_t^n) is progressively measurable and a.s. continuous, valued in $\mathbb{R}^d$. If

$$\lim_{m\to\infty} \sup_{n\geqslant m} \mathbb{P}\left[\sup_{0\leqslant t\leqslant T} |X_t^n - X_t^m| \geqslant \varepsilon\right] = 0,$$

then there exists a progressively measurable and a.s. continuous (X_t), such that

$$\lim_{m\to\infty} \sup_{n\geqslant m} \mathbb{P}\left[\sup_{0\leqslant t\leqslant T} |X_t^n - X_t| \geqslant \varepsilon\right] = 0, \qquad T > 0, \ \varepsilon > 0.$$

The lemma is a kind of stochastic completeness, whose proof is long and tedious, which we omit it here. Refer to [63; Lemma 4.3.3].

Till now, we have proved: there exists progressively measurable and a.s. continuous (M_t) such that

$$\lim_{m\to\infty} \sup_{n\geqslant m} \mathbb{P}\left[\sup_{0\leqslant t\leqslant T} |M_t^n - M_t| \geqslant \varepsilon\right] = 0, \qquad T > 0, \ \varepsilon > 0.$$

In particular, for each fixed t, $M_t^n \xrightarrow{\mathbb{P}} M_t$ as $n \to \infty$. The above proof indeed gives that $M_t^n \xrightarrow{L^1(\mathbb{P})} M_t$. Therefore $(M_t, \mathscr{F}_t, \mathbb{P})$ is an a.s. continuous martingale.

Finally, let $A' = X^2 - 2M$. Then A' is right-continuous and progressively measurable. On the other hand, from d) above we see

$$\lim_{m\to\infty} \sup_{n\geqslant m} \mathbb{P}\left[\sup_{0\leqslant t\leqslant T} |A_t^n - A_t'| \geqslant \varepsilon\right] = 0, \qquad T > 0, \ \varepsilon > 0.$$

Further from a) and c) above we know

$$A_0' = 0, \qquad A_t' \text{ non-decreasing}, \qquad \text{a.s.}$$

Thus, we just choose

$$A_0 \equiv 0, \qquad A_t = \sup_{0\leqslant s<t} (A_s' - A_0'), \qquad t > 0,$$

to derive the assertion as required. □

Definition 8.6. Let $\mathscr{M}_c^2$ be the total of a.s. continuous $L^2(\mathbb{P})$ martingales $(X_t, \mathscr{F}_t, \mathbb{P})$. For $X = (X_t) \in \mathscr{M}_c^2$, let $\langle X \rangle = (\langle X \rangle(t))$ be the increasing process in the previous decomposition theorem, which is called the **quadratic variation process**. And

$$\langle X, Y \rangle = \frac{1}{4}(\langle X + Y \rangle - \langle X - Y \rangle), \qquad X, Y \in \mathscr{M}_c^2$$

is called **jointed quadratic variation process**.

Lemma 8.7. The process $\langle X, Y \rangle$ is progressively measurable and a.s. continuous with locally bounded variation. Furthermore $\langle X, Y \rangle(T) \in L^1(\mathbb{P})$ for each $T \geqslant 0$.

Proof Since

$$\langle X \rangle(T) = |A|(T) = A_T = X_T^2 - M_T,$$

we have

$$\mathbb{E}|A|(T) = \mathbb{E}A_T = \mathbb{E}X_T^2 < \infty.$$

Therefore $\langle X \pm Y \rangle(T) \in L^1(\mathbb{P})$.   □

8.3   Properties of Variation Processes

Theorem 8.8. For $X, Y \in \mathscr{M}_c^2$, $\langle X, Y \rangle$ is the unique a.s. continuous and progressively measurable process with locally bounded variation, such that $\langle X, Y \rangle(0) \equiv 0$ and

$$(X_t Y_t - \langle X, Y \rangle(t), \ \mathscr{F}_t, \ \mathbb{P})$$

is a martingale. In particular, for $X, Y, Z \in \mathscr{M}_c^2$, we have
  (1) $\langle X \rangle = \langle X, X \rangle$, a.s.
  (2) Linearity: $\langle aX + bY, Z \rangle = a\langle X, Z \rangle + b\langle Y, Z \rangle$, a.s.
  (3) Schwarz's inequality: $|\langle X, Y \rangle|(\Gamma) \leqslant \langle X \rangle(\Gamma)^{1/2} \langle Y \rangle(\Gamma)^{1/2}$ a.s. $\Gamma \in \mathscr{B}([0, \infty])$; (therefore $|\langle X, Y \rangle|(\Gamma) \leqslant \frac{1}{2}(\langle X \rangle(\Gamma) + \langle Y \rangle(\Gamma)))$
  (4) Minkowski's inequality: $\langle X + Y \rangle(\Gamma) \leqslant \langle X \rangle(\Gamma)^{1/2} + \langle Y \rangle(\Gamma)^{1/2}$, a.s.

Proof The first assertion is just Lemma 8.7. We only prove (3), the others can be deduced from the uniqueness or proved similarly. For this, we need only check

$$|\langle X, Y \rangle(t) - \langle X, Y \rangle(s)| \leqslant (\langle X \rangle(t) - \langle X \rangle(s))^{1/2}(\langle Y \rangle(t) - \langle Y \rangle(s))^{1/2}, \qquad \text{a.s.}$$

Replace X and Y by ${}^s X$ and ${}^s Y$, respectively, the proof is reduced to

$$|\langle X, Y \rangle(t)| \leqslant \langle X \rangle(t)^{1/2} \langle Y \rangle(t)^{1/2}, \qquad \text{a.s.} \tag{8.2}$$

Its proof can be completed by the usual argument for the relation between inner product and norm. By linearity, for $\lambda \neq 0$,

$$0 \leqslant \langle \lambda X \pm Y/\lambda \rangle(t) = \lambda^2 \langle X \rangle(t) \pm 2\langle X, Y \rangle(t) + \langle Y \rangle(t)/\lambda^2, \qquad \text{a.s.},$$

which implies

$$|\langle X, Y \rangle(t)| \leqslant \frac{1}{2}[\lambda^2 \langle X \rangle(t) + \langle Y \rangle(t)/\lambda^2], \qquad \text{a.s.}$$

Take $\lambda^2 = (\langle Y \rangle(t)/\langle X \rangle(t))^{1/2}$ to get the minimum in the right-hand side. This proves (8.2).   □

8.4   Stochastic Integral

For a progressively measurable, increasing and a.s. continuous process $A = (A_t)$, let

$$L^2_{\ell oc}(A, \mathbb{P}) = \Big\{ \alpha : [0, \infty) \times \Omega \to \mathbb{R} \text{ is progressively measurable,}$$

$$\text{and } \mathbb{E} \int_0^T \alpha_t^2 A(\mathrm{d}t) < \infty \text{ for each } T > 0 \Big\}.$$

For given $X \in \mathcal{M}_c^2$, $\alpha \in L^2_{\ell oc}(\langle X \rangle, \mathbb{P})$, we will prove there exists a unique $I : [0, \infty) \times \Omega \to \mathbb{R}$ such that

(1) $I(0) = 0$, $I \in \mathcal{M}_c^2$;
(2) $\langle I, Y \rangle(\mathrm{d}t) = \alpha \langle X, Y \rangle(\mathrm{d}t)$, a.s. for each $Y \in \mathcal{M}_c^2$.

Proof of uniqueness   Suppose there exist two I and I' satisfying the above conditions. Then for all $Y \in \mathcal{M}_c^2$, $\langle I - I', Y \rangle \equiv 0$, a.s. In particular, by taking $Y = I - I'$, $\langle I - I' \rangle = 0$, a.s. Hence

$$\mathbb{E}[I(T) - I'(T)]^2 = \mathbb{E}[M_T] + \mathbb{E}\langle I - I' \rangle(T) = \mathbb{E}M_T = \mathbb{E}(M_0) = 0.$$

Therefore $I = I'$, a.s. $\square$

Definition 8.9. The previous I is called the **Itô's stochastic integral** of α with respect to X, and is denoted by $I = \alpha_\bullet X = \int_0^\bullet \alpha_t \mathrm{d}X_t$.

Different from Chapter 7, the definition here is axiomatic rather than constructive.

Lemma 8.10. If $I_\alpha = \alpha_\bullet X$ exists, then $\mathbb{E}[I_\alpha(T)^2] = \mathbb{E}\Big[\int_0^T \alpha_t^2 \langle X \rangle(\mathrm{d}t) \Big]$.

Proof   From Condition (2), we have $\langle I_\alpha, Y \rangle = \alpha \langle X, Y \rangle$ for each $Y \in \mathcal{M}_c^2$. Thus $\langle I_\alpha, I_\alpha \rangle = \alpha \langle X, I_\alpha \rangle = \alpha^2 \langle X \rangle$. Furthermore

$$\mathbb{E}\Big[\Big(\int_0^T \alpha_t \mathrm{d}X_t \Big)^2 \Big] = \mathbb{E}[M(T)] + \mathbb{E}[\langle I_\alpha, I_\alpha \rangle(T)]$$

$$= \mathbb{E}\Big[\int_0^T \alpha_t^2 \langle X \rangle(\mathrm{d}t) \Big]. \quad \square$$

Lemma 8.11 (Linearity). Assume that $\alpha_\bullet X$ and $\beta_\bullet X$ exist for $\alpha, \beta \in L^2_{\ell oc}(\langle X \rangle, \mathbb{P})$. Then $(a\alpha + b\beta)_\bullet X$ also exists and

$$(a\alpha + b\beta)_\bullet X = a(\alpha_\bullet X) + b(\beta_\bullet X), \qquad a, b \in \mathbb{R}.$$

Furthermore

$$\mathbb{E}\Big[\sup_{0 \leqslant t \leqslant T} \big((\alpha_\bullet X)_t - (\beta_\bullet X)_t \big)^2 \Big] \leqslant 4\mathbb{E}\Big[\int_0^T (\alpha_t - \beta_t)^2 \langle X \rangle(\mathrm{d}t) \Big], \quad T > 0.$$

Proof We only prove the latter assertion. By Doob's inequality

$$\mathbb{E}\left[\sup_{0\leqslant t\leqslant T}((\alpha_\bullet X)_t - (\beta_\bullet X)_t)^2\right] \leqslant 4\mathbb{E}\left[((\alpha_\bullet X)_T - (\beta_\bullet X)_T)^2\right]$$

$$= 4\mathbb{E}\left[\left(\int_0^T \alpha_t dX_t - \int_0^T \beta_t dX_t\right)^2\right]$$

$$= 4\mathbb{E}\left[\left(\int_0^T (\alpha_t - \beta_t) dX_t\right)^2\right] \quad \text{(linearity)}$$

$$= 4\mathbb{E}\left[\int_0^T (\alpha_t - \beta_t)^2 \langle X\rangle(dt)\right].$$

Here in the last step, we have used Lemma 8.10. $\square$

Now we turn to study the existence of stochastic integral. We go back the construction argument as in Chapter 7. We begin with a special case.

Theorem 8.12. $\alpha_\bullet X$ exists for each $\alpha \in L^2_{\ell oc}(\langle X\rangle, \mathbb{P})$.

Proof We consider first the simple process α: there exists $n \geqslant 1$ such that

$$\alpha_t = \alpha_{[nt]/n}, \quad t \geqslant 0.$$

Let

$$I(t) = \sum_{k=0}^{\infty} \alpha_{k/n}\left[X_{t\wedge\frac{k+1}{n}} - X_{t\wedge\frac{k}{n}}\right].$$

Then $I \in \mathcal{M}_c^2$. Further, if $k/n \leqslant s \leqslant (k+1)/n$, then for $Y \in \mathcal{M}_c^2$,

$$\mathbb{E}[I(t)Y_t - I(s)Y_s|\mathscr{F}_s] = \mathbb{E}[(I(t) - I(s))(Y_t - Y_s)|\mathscr{F}_s]$$

$$= \alpha_{k/n}\mathbb{E}[(X_t - X_s)(Y_t - Y_s)|\mathscr{F}_s]$$

$$= \alpha_{k/n}\mathbb{E}[X_tY_t - X_sY_s|\mathscr{F}_s]$$

$$= \mathbb{E}[(\alpha\langle X, Y\rangle)(t) - (\alpha\langle X, Y\rangle)(s)|\mathscr{F}_s], \quad \text{a.s.}$$

In other words,

$$\langle I, Y\rangle = \alpha\langle X, Y\rangle, \quad u.s., \quad Y \in \mathcal{M}_c^2.$$

Thus $\alpha_\bullet X$ exists and equals I defined as above.

Now suppose α is bounded, progressively measurable and a.s. continuous. Take

$$\alpha_t^{(n)} = \alpha_{[nt]/n}.$$

Then $\alpha^{(n)} \xrightarrow{\text{a.s.}} \alpha$. By Lemma 8.11 and the dominated convergence theorem, we have

$$\mathbb{E}\left[\sup_{0\leqslant t\leqslant T}\left(\int_0^t \alpha_u^{(n)} dX_u - \int_0^t \alpha_u^{(m)} dX_u\right)^2\right]$$

$$\leqslant 4\mathbb{E}\left[\int_0^T (\alpha_t^{(n)} - \alpha_t)^2 \langle X\rangle(dt)\right] \to 0, \quad n \to \infty.$$

Therefore, we can define $\int_0^\bullet \alpha_u dX_u$ as the limit of $\int_0^\bullet \alpha_u^{(n)} dX_u$. We have

$$\mathbb{E}\left[\sup_{0 \leqslant t \leqslant T} \left(\int_0^t \alpha_u^{(n)} dX_u - \int_0^t \alpha_u dX_u \right)^2 \right]$$

$$\leqslant 4\mathbb{E}\left[\int_0^T (\alpha_t^{(n)} - \alpha_t)^2 \langle X \rangle(dt) \right] \to 0, \qquad n \to \infty.$$

To complete the construction for the general case of $\alpha_\bullet X$, we need only use the following approximation.

Lemma 8.13. Assume $A : [0, \infty) \times \Omega \to [0, \infty)$ is increasing, a.s. continuous and progressively measurable such that $A(0) \equiv 0$. For $\alpha \in L^2_{\ell oc}(A, \mathbb{P})$, there exists a sequence of bounded and a.s. continuous $\{\alpha^{(n)}\} \subset L^2_{\ell oc}(A, \mathbb{P})$, such that

$$\alpha^{(n)} \xrightarrow{L^2_{\ell oc}(A, \mathbb{P})} \alpha.$$

Proof Since the set of bounded elements in $L^2_{\ell oc}(A, \mathbb{P})$ is obviously dense in $L^2_{\ell oc}(A, \mathbb{P})$, we may and do assume α is bounded.

Consider first the simple case that $A_t \equiv t$ ($t \geqslant 0$). Take $\rho \in C_0^\infty(0, 1)$ such that $\int \rho(t) dt = 1$ (on $\mathbb{R}^d$, the mollifier ρ can be defined by:

$$\rho(x) = \begin{cases} c \exp\left[\frac{1}{|x|^2 - 1} \right], & \text{if } |x| < 1; \\ 0, & \text{if } |x| \geqslant 1, \end{cases}$$

where c is the normalization constant). Notice $\operatorname{supp} \rho$ is contained in the unit ball centered at the origin. Extend α to a function on $\mathbb{R} \times \Omega$: $\alpha_t \equiv 0$ if $t < 0$. Define $\alpha_t^{(n)} = n \int \alpha_{t-s} \rho(ns) ds$ for every $n \geqslant 1$. It is easy to prove that $\alpha^{(n)}$ is what we require. $\square$

Next consider the general case. We need to prove: for each $T > 0$ and $\varepsilon > 0$, there exists a bounded and a.s. continuous $\bar{\alpha} \in L^2_{\ell oc}(A, \mathbb{P})$ such that

$$\mathbb{E}\left[\int_0^T (\alpha_t - \bar{\alpha}_t)^2 A(dt)) \right] < \varepsilon^2.$$

For $T, \varepsilon > 0$, take $M > 1$ such that

$$\mathbb{E}\left[\int_0^T \alpha_t^2 A(dt); \, A_T \geqslant M - 1 \right] < \left(\frac{\varepsilon}{2} \right)^2.$$

Next, take $\eta \in C^\infty(\mathbb{R})$ such that

$$I_{[0, M-1]} \leqslant \eta \leqslant I_{[-1, M]}.$$

Let

$$D(t) = \int_0^t \eta(A_s)^2 A(ds) + t, \qquad \tau(t) = D^{-1}(t) \text{ (inverse function)}, \qquad t \geqslant 0.$$

(In the definition of $D(t)$, we add the term t to guarantee $D(t)$ to be strictly increasing, then to ensure the monotonicity of the inverse function.) Then $\{\tau_t\}_{t\geqslant 0}$ is a family of increasing bounded stopping times. Set

$$\mathscr{G}_t = \mathscr{F}_{\tau(t)}, \qquad \beta_t = \alpha_{\tau(t)}. \quad \text{(time change)}$$

Since β is bounded and $(\mathscr{G}_t)$-progressively measurable, from the previous proof, we know that there exists bounded continuous and $(\mathscr{G}_t)$-progressively measurable $\bar{\beta}$ such that

$$\mathbb{E}\left[\int_0^{T+M} \left(\beta_t - \bar{\beta}_t\right)^2 dt\right] < \left(\frac{\varepsilon}{2}\right)^2.$$

Finally, let $\bar{\alpha}_t = \bar{\beta}_{D(t)}\eta(A_t)$. Then $\bar{\alpha}$ is bounded and a.s. continuous element in $L^2_{\ell oc}(A, \mathbb{P})$. Thus

$$\mathbb{E}\left[\int_0^T (\alpha_t - \bar{\alpha}_t)^2 A(dt))\right]^{1/2}$$

$$= \mathbb{E}\left[\int_0^T \left(\alpha_t - \bar{\beta}_{D(t)}\eta(A_t)\right)^2 A(dt)\right]^{1/2}$$

$$\leqslant \mathbb{E}\left[\int_0^T \alpha_t^2\left(1 - \eta(A_t)\right)^2 A(dt)\right]^{1/2}$$

$$\qquad + \mathbb{E}\left[\int_0^T \left(\alpha_t - \bar{\beta}_{D(t)}\right)^2 \eta(A_t)^2 A(dt))\right]^{1/2}$$

$$\leqslant \frac{\varepsilon}{2} + \mathbb{E}\left[\int_0^T \left(\alpha_t - \bar{\beta}_{D(t)}\right)^2 D(dt)\right]^{1/2}$$

$$= \frac{\varepsilon}{2} + \mathbb{E}\left[\int_0^T \left(\beta_{D(t)} - \bar{\beta}_{D(t)}\right)^2 D(dt)\right]^{1/2}$$

$$\left(\text{as } \alpha_t = \alpha_{D^{-1}(D(t))} = \alpha_{\tau(D(t))} = \beta_{D(t)}\right)$$

$$\leqslant \frac{\varepsilon}{2} + \mathbb{E}\left[\int_0^{T+M} (\beta_t - \bar{\beta}_t)^2 dt\right]^{1/2}$$

$$< \frac{\varepsilon}{2} + \frac{\varepsilon}{2} = \varepsilon. \quad \square$$

8.5 Itô's Formula

The following Itô's formula is a generalization of Theorem 7.6.

Theorem 8.14 (Itô's formula). Assume that $X = (X^1, \cdots, X^M) \in (\mathscr{M}_c^2)^M$ and $Y : [0, \infty) \times \Omega \to \mathbb{R}^N$ is a.s. continuous and progressively measurable with locally bounded variation:

$$|Y|(T) \equiv \left(\sum_1^N |Y^j|(T)^2 \right)^{1/2} \in L^1(\mathbb{P}), \quad T > 0.$$

Let $Z = (X, Y)$. Then for every $f \in C_b^{2,1}(\mathbb{R}^M \times \mathbb{R}^N)$, we have

$$f(Z_T) - f(Z_0) = \sum_{i=1}^M \int_0^T \partial_i^x f(Z_t) \mathrm{d}X_t^i$$

(integral of the first order derivative w.r.t. martingale)

$$+ \sum_{j=1}^N \int_0^T \partial_j^y f(Z_t) Y^j(\mathrm{d}t)$$

(integral of the first order derivative w.r.t. the variation process)

$$+ \frac{1}{2} \sum_{i,j=1}^M \int_0^T \partial_{ij}^x f(Z_t) \langle X^i, X^j \rangle(\mathrm{d}t), \text{ a.s.}$$

(integral of the second derivative w.r.t. quadratic variation process)

Remark 8.15. When $Y \equiv 0$, the second sum on in the right-hand side disappears. We further consider the deterministic case of $M = 1$, then

$$f(X_T) - f(X_0) = \int_{X_0}^{X_T} \frac{\mathrm{d}f}{\mathrm{d}x}(x)\mathrm{d}x = \int_0^T \frac{\mathrm{d}f}{\mathrm{d}x}(X_t)\mathrm{d}X_t.$$

This is the basic formula in classical calculus. In the stochastic case, there is an additional quadratic term $\langle , \rangle$. In other words, in stochastic calculus, the stochastic process can not be approximated by the first order derivative. It should use the second derivative. Accordingly, when studying stochastic differential geometry, the first order vector fields are insufficient and we have to use the second vector fields. Thus the stochastic differential geometry is called "the second order differential geometry", while the deterministic case is called "the first order differential geometry".

As a basic formula in stochastic calculus, Itô's formula has numerous applications. We give one example here. Assume $B_t = (B_t^1, \cdots, B_t^d)$ is d-dimensional BM^d. Then by Itô's formula,

$$f(B_t) - f(B_0) = \sum_{i=1}^d \int_0^t \partial_i^x f(B_s)\mathrm{d}B_s + \frac{1}{2} \sum_{i,j=1}^d \int_0^t \partial_{ij}^x f(B_s)\langle B^i, B^j \rangle(\mathrm{d}s)$$

for each $f \in C_0^\infty(\mathbb{R}^d)$. Note

$$\langle B^i, B^j \rangle = 0, \quad i \neq j. \quad \text{(by independence)}$$

Thus

$$f(B_t) - f(B_0) = \sum_{i=1}^{d} \int_0^t \partial_i^x f(B_s) \mathrm{d}B_s + \frac{1}{2} \int_0^t \Delta f(B_s) \mathrm{d}s.$$

It follows that $\left(f(B_t) - \frac{1}{2}\int_0^t \Delta f(B_s)\mathrm{d}s, \mathscr{F}_t, \mathbb{P}\right)$ is a martingale. This a martingale description of BM^d which was mentioned in Section 6.4 without proof.

Proof of Itô's formula. Denote $\langle\!\langle X, X\rangle\!\rangle(t, \omega) = (\langle X^i, X^j\rangle(t, \omega) : 1 \leqslant i, j \leqslant M)$. Without loss of generality assume $t \to Z_t(\omega)$, $t \to \langle\!\langle X, X\rangle\!\rangle(t, \omega)$ is continuous for every $\omega \in \Omega$, and $f \in C_b^\infty(\mathbb{R}^{N+N})$. Given $n \geqslant 1$, define a sequence of stopping times τ_k^n as follows:

$$\tau_0^n \equiv 0,$$

$$\tau_{k+1}^n = \inf\left\{ t \geqslant \tau_k^n : \left(\max_{1 \leqslant i \leqslant M} (\langle X^i\rangle(t) - \langle X^i\rangle(\tau_k^n))\right) \vee |Z_t - Z_{\tau_k^n}| \geqslant \frac{1}{n}\right\}$$

$$\bigwedge\left(\tau_k^n + \frac{1}{n}\right)\bigwedge T.$$

Then $\tau_k^n = T$ for each $T > 0$ and $\omega \in \Omega$, except finitely many k's. Hence,

$$f(Z_T) - f(Z_0) = \sum_{k=0}^{\infty} [f(Z_{k+1}^n) - f(Z_k^n)],$$

where

$$Z_k^n = (X_k^n, Y_k^n) := Z_{\tau_k^n}.$$

Notice

$$g(Y_{k+1}^n) - g(Y_k^n) = \sum_{j=1}^{N} \int_{\tau_k^n}^{\tau_{k+1}^n} \partial_j^y g(Y_t) Y^j(\mathrm{d}t).$$

Indeed, since

$$\frac{\mathrm{d}g}{\mathrm{d}t}(Y_t) = \sum_{j=1}^{N} \partial_j g(Y_t) \frac{\mathrm{d}Y^j}{\mathrm{d}t},$$

we have

$$g(Y_{t_2}) - g(Y_{t_1}) = \sum_{j=1}^{N} \int_{t_1}^{t_2} \partial_j g(Y_t)\left(\frac{\mathrm{d}Y^j}{\mathrm{d}t}\right)\mathrm{d}t = \sum_{j=1}^{N} \int_{t_1}^{t_2} \partial_j g(Y_t) Y^j(\mathrm{d}t),$$

where we use the fact that Y_t^j has bounded variation.

Therefore, we obtain the following decomposition:

$$f(Z_{k+1}^n) - f(Z_k^n)$$
$$= [f(X_{k+1}^n, Y_k^n) - f(X_k^n, Y_k^n)] + [f(X_{k+1}^n, Y_{k+1}^n) - f(X_{k+1}^n, Y_k^n)]$$
$$= \sum_{i=1}^M \partial_i^x f(Z_k^n) \Delta_k^n X^i + \frac{1}{2} \sum_{i,j=1}^M \partial_{ij}^x f(Z_k^n) \Delta_k^n \langle X^i, X^j \rangle$$
$$+ \sum_{j=1}^N \int_{\tau_k^n}^{\tau_{k+1}^n} \partial_j^y f(X_{k+1}^n, Y_t) Y^j(\mathrm{d}t) + R_k^n,$$

where $\Delta_k^n \xi = \xi(\tau_{k+1}^n) - \xi(\tau_k^n)$, and

$$R_k^n = \frac{1}{2} \sum_{i,j=1}^M [\partial_{ij}^x f(\hat{Z}_k^n) - \partial_{ij}^x f(Z_k^n)] \Delta_k^n X^i \Delta_k^n X^j$$

$$+ \frac{1}{2} \sum_{i,j=1}^M [\partial_{ij}^x f(Z_k^n)(\Delta_k^n X^i \Delta_k^n X^j - \Delta_k^n \langle X^i, X^j \rangle),$$

and $\hat{Z}_k^n$ is a point on the line segment connecting (X_{k+1}^n, Y_k^n) and (X_k^n, Y_k^n).
For the stochastic integral of an elementary function, we have

$$\sum_{k=0}^\infty \partial_i^x f(Z_k^n) \Delta_k^n X^i = \int_0^T \partial_i^x f(Z_t^n) \mathrm{d}X_t^i,$$

where

$$Z_t^n = \begin{cases} Z_k^n, & \text{if } t \in [\tau_k^n, \tau_{k+1}^n); \\ Z_T, & \text{if } t \geqslant T. \end{cases}$$

Since $Z_t^n \to Z_t$ uniformly in $t \in [0, T]$, we have

$$\sum_{k=0}^\infty \partial_i^x f(Z_k^n) \Delta_k^n X^i \xrightarrow{L^2(\mathbb{P})} \int_0^T \partial_i^x f(Z_t) \mathrm{d}X_t^i.$$

On the other hand, according to the usual calculus, we have

$$\sum_{k=0}^\infty \partial_{ij}^x f(Z_k^n) \Delta_k^n \langle X^i, X^j \rangle \xrightarrow{L^1(\mathbb{P})} \int_0^T \partial_{ij}^x f(Z_t) \langle X^i, X^j \rangle (\mathrm{d}t),$$

$$\sum_{k=0}^\infty \int_{\tau_k^n}^{\tau_{k+1}^n} \partial_j^y f(X_{k+1}^n, Y_t) Y^j(\mathrm{d}t) \xrightarrow{L^1(\mathbb{P})} \int_0^T \partial_j^y f(Z_t) Y^j(\mathrm{d}t).$$

To complete the proof, we need only prove $\sum_k R_k^n \xrightarrow{\mathbb{P}} 0$. Firstly,

$$|[\partial_{ij}^x f(\hat{Z}_k^n) - \partial_{ij}^x f(Z_k^n)] \Delta_k^n X^i \Delta_k^n X^j|$$
$$\leqslant C[(\Delta_k^n X^i)^2 + (\Delta_k^n X^j)^2] |\hat{Z}_k^n - Z_k^n|$$
$$\leqslant C'[(\Delta_k^n X^i)^2 + (\Delta_k^n X^j)^2]/n.$$

Thus,

$$\mathbb{E}\left\{\left|\sum_k \left[(\partial_{ij}^x f(\hat{Z}_k^n) - \partial_{ij}^x f(Z_k^n))\Delta_k^n X^i \Delta_k^n X^j\right]\right|\right\}$$

$$\leqslant 2C''\mathbb{E}\left[|X_T - X_0|^2\right]/n \quad \text{(by orthogonality)}$$

$$\to 0.$$

Secondly,

$$\mathbb{E}\left[\left(\sum_{k=0}^{\infty} \partial_{ij}^x f(Z_k^n)\left(\Delta_k^n X^i \Delta_k^n X^j - \Delta_k^n \langle X^i, X^j\rangle\right)\right)^2\right]$$

$$= \sum_{k=0}^{\infty} \mathbb{E}\left[\left(\partial_{ij}^x f(Z_k^n)(\Delta_k^n X^i \Delta_k^n X^j - \Delta_k^n \langle X^i, X^j\rangle)\right)^2\right]$$

$$\leqslant C \sum_{k=0}^{\infty} \mathbb{E}\left[\left(\Delta_k^n X^i \Delta_k^n X^j - \Delta_k^n \langle X^i, X^j\rangle\right)^2\right]$$

$$\leqslant C' \sum_{k=0}^{\infty} \mathbb{E}\left[\left(\Delta_k^n X^i\right)^4 + \left(\Delta_k^n X^j\right)^4 + \left(\Delta_k^n \langle X^i\rangle\right)^2 + \left(\Delta_k^n \langle X^j\rangle\right)^2\right]$$

$$\text{(Schwarz's inequality)}$$

$$\leqslant C' \sum_{k=0}^{\infty} \mathbb{E}\left[\left(\Delta_k^n X^i\right)^2/n^2 + \left(\Delta_k^n X^j\right)^2/n^2 + \Delta_k^n \langle X^i\rangle/n + \Delta_k^n \langle X^j\rangle/n\right]$$

$$\text{(by the definition of } \tau_k^n)$$

$$\leqslant C' \sum_{k=0}^{\infty} \mathbb{E}\left[\left(\Delta_k^n X^i\right)^2 + \left(\Delta_k^n X^j\right)^2 + \Delta_k^n \langle X^i\rangle + \Delta_k^n \langle X^j\rangle\right]/n$$

$$\leqslant C'' \sum_{k=0}^{\infty} \mathbb{E}\left[|\Delta_k^n X|^2\right]/n \quad \text{(by martingale property)}$$

$$= C''\mathbb{E}\left[|X_T - X_0|^2\right]/n \quad \text{(by orthogonality)}$$

$$\to 0, \qquad n \to \infty.$$

In the last inequality, we have used the decomposition: $\left|\Delta_k^n X\right|^2 = \sum_{i=1}^{M} \left(\Delta_k^n X^i\right)^2$. $\square$

8.6   Local Martingale and Semimartingale

Till now, we have studied the following three problems for square integrable martingale (X_t).

(1) Doob-Meyer decomposition: $X_t^2 = M_t$ (martingale) $+ \langle X \rangle(t)$ (locally bounded variation).

(2) For $\alpha \in L_{\ell oc}^2 (\langle X \rangle)$, we have defined its stochastic integral and stochastic differential.

(3) Itô's formula, the differential formula for composed function: for

$$f \in C_b^{2,1}(\mathbb{R}^M \otimes \mathbb{R}^N) \quad \text{and} \quad Z = (X, Y),$$

it holds that

$$df(Z_t) = \sum_{i=1}^{M} \partial_i^x f(Z_t) dX_t^i \sum_{j=1}^{N} \partial_j^y f(Z_t) Y^j(dt)$$

$$+ \frac{1}{2} \sum_{i,j=1}^{M} \partial_{ij}^x f(Z_t) \langle X^i, X^j \rangle(dt).$$

However the condition of square integrability is somewhat too restrictive; thus, we relax it.

Definition 8.16. Assume (X_t) is progressively measurable and a.s. continuous. If there exists a sequence of stopping time $\sigma_n \uparrow \infty$, a.s. such that for each $n \geqslant 1$, $(X_t^{\sigma_n} := X_{t \wedge \sigma_n})_{t \geqslant 0}$ is a bounded martingale, then $(X_t)_{t \geqslant 0}$ is called an a.s. continuous **local martingale**. The total of local martingales is denoted by $\mathcal{M}_c^{\ell oc}$.

Assume $X \in \mathcal{M}_c^{\ell oc}$. Then we also have the Doob-Meyer decomposition

$$X_t^2 = M_t \left(\in \mathcal{M}_c^{\ell oc} \right) + \langle X \rangle(t) \quad \text{(also locally bounded variation)}.$$

Indeed, since

$$(X_t^{\sigma_n})^2 = M_t^{\sigma_n} + \langle X^{\sigma_n} \rangle(t),$$

we need only take $\langle X \rangle_t = \sup_n \langle X^{\sigma_n} \rangle(t)$. From this we deduce the existence of the decomposition. Following the previous proof, we can prove the uniqueness of the decomposition. And then define $\langle X, Y \rangle$. Next, we can define the stochastic integral for $\alpha \in L_{\ell oc}^2(\langle X \rangle)$. Finally, the condition $f \in C_b^{2,1}$ in Itô's formula can be relaxed to $f \in C^{2,1}$.

Let us turn to another problem: which set of processes is closed under the operation of change of variable? We have already known

$$df(X_t) = f'(X_t) dX_t \quad \text{(local martingale)}$$

$$+ \frac{1}{2} f''(X_t) \langle X \rangle(dt) \quad \text{(locally bounded variation)}$$

for every $X \in \mathcal{M}_c^{\ell oc}$. The right-hand side indicates that it should contain two parts.

Definition 8.17. $Z = (Z_t)$ is called an a.s. continuous **semimartingale**, if Z has the decomposition: $Z = X + Y$, where $X \in \mathcal{M}_c^{\ell oc}$, Y is progressively measurable and a.s. continuous with locally bounded variation. We denote the total of a.s. continuous semimartingales by $\mathcal{S}_\bullet \mathcal{M}_c$.

Remark 8.18. The composition (X, Y) is a.s. unique for $Z \in \mathcal{S}_\bullet \mathcal{M}_c$. Therefore, denote $\langle Z \rangle = Y$. Similarly, for $Z, Z' \in \mathcal{S}_\bullet \mathcal{M}_c$, one may define $\langle Z, Z' \rangle$. Furthermore, for the progressively measurable process α satisfying

$$\int_0^T \alpha_t^2 \langle Z \rangle (\mathrm{d}t) \bigvee \int_0^T |\alpha_t||Y|(\mathrm{d}t) < \infty, \qquad T > 0, \text{ a.s.}$$

we can define the stochastic integral

$$\int_0^\bullet \alpha_s \mathrm{d}Z_s = \int_0^\bullet \alpha_s \mathrm{d}X_s + \int_0^\bullet \alpha_s \mathrm{d}Y_s.$$

And Itô's formula becomes:

$$f(Z_t) - f(Z_0) = \sum_{i=1}^N \int_0^t \partial_i^x f(Z_s) \mathrm{d}Z_s^i + \frac{1}{2} \sum_{i,j=1}^N \int_0^t \partial_{ij}^x f(Z_s) \langle Z^i, Z^j \rangle (\mathrm{d}s), \text{ a.s.}$$

$$Z \in (\mathcal{S}_\bullet \mathcal{M}_c)^N, \quad f \in C^2(\mathbb{R}^N).$$

8.7   Multivariate Stochastic Integral

Assume $X = (X^1, \cdots, X^d) \in (\mathcal{M}_c^2)^d$. Denote

$$\langle\!\langle X, X \rangle\!\rangle = (\langle X^i, X^j \rangle : 1 \leqslant i, j \leqslant d).$$

Let

$$L_{\ell oc}^2(\langle\!\langle X, X \rangle\!\rangle) = \Big\{ \theta : [0, \infty) \times \Omega \to \mathbb{R}^d : \theta \text{ progressively measurable and}$$

$$\mathbb{E}\left[\int_0^T (\theta_t, \langle\!\langle X, X \rangle\!\rangle (\mathrm{d}t) \theta_t)_{\mathbb{R}^d} \right]$$

$$= \mathbb{E}\left[\int_0^T \theta_t^* \langle\!\langle X, X \rangle\!\rangle (\mathrm{d}t) \theta_t \right]$$

$$= \sum_{i,j=1}^d \mathbb{E}\left[\int_0^T \theta_t^i \langle X^i, X^j \rangle (\mathrm{d}t) \theta_t^j \right] < \infty \text{ for every } T > 0 \Big\},$$

where θ^* denotes the transpose of θ.

When dealing with multivariate case, it is natural to use matrices and vectors. Note

$$|\theta_t^i \langle X^i, X^j \rangle (\mathrm{d}t) \theta_t^j| \leqslant |\theta_t^i \theta_t^j| \langle X^i \rangle^{1/2} (\mathrm{d}t) \langle X^j \rangle^{1/2} (\mathrm{d}t)$$

$$\leqslant \frac{1}{2} |\theta_t^i \theta_t^j| (\langle X^i \rangle (\mathrm{d}t) + \langle X^j \rangle (\mathrm{d}t)).$$

Thus $\langle\!\langle X, X\rangle\!\rangle$ is nonnegative definite. Moreover $\int_0^T \theta_t^* \langle\!\langle X, X\rangle\!\rangle(\mathrm{d}t)\theta_t \geqslant 0$, a.s. For $\theta, \theta' \in L^2_{\ell oc}(\langle\!\langle X, X\rangle\!\rangle)$, we see

$$\langle\theta, \theta'\rangle \equiv \mathbb{E}\left[\int_0^T \theta_t^* \langle\!\langle X, X\rangle\!\rangle(\mathrm{d}t)\theta_t\right]$$

becomes the inner product on $L^2_{\ell oc}(\langle\!\langle X, X\rangle\!\rangle)$ (locally, that is for fixed T).

Remark 8.19. Locally, $\left(L^2_{\ell oc}(\mathrm{Tr}\langle\!\langle X, X\rangle\!\rangle)\right)^d$ can be viewed as a dense subspace of $L^2_{\ell oc}(\langle\!\langle X, X\rangle\!\rangle)$. Indeed, for $\theta \in L^2_{\ell oc}(\langle\!\langle X, X\rangle\!\rangle)$, let $\theta_t(n) = I_{[0,n]}(|\theta_t|)\theta_t$. Then for each i,

$$\mathbb{E}\left[\int_0^T \theta_t^i(n)^2 \mathrm{Tr}\langle\!\langle X, X\rangle\!\rangle(\mathrm{d}t)\right] \leqslant n^2 \sum_{i=1}^d \mathbb{E}[\langle X^i\rangle(T)] < \infty.$$

Therefore $\theta(n) \in \left(L^2_{\ell oc}(\mathrm{Tr}\langle\!\langle X, X\rangle\!\rangle)\right)^d$. Moreover, by the pointwise convergence and the dominated convergence theorem,

$$\|\theta - \theta(n)\| := \langle\theta - \theta(n), \theta - \theta(n)\rangle$$
$$= \mathbb{E}\left[\int_0^T (\theta - \theta(n))_t^* \langle\!\langle X, X\rangle\!\rangle(\mathrm{d}t)(\theta - \theta(n))_t\right]$$
$$\to 0, \qquad n \to \infty.$$

Define

$$\int_0^T \theta_t \mathrm{d}X_t = \int_0^T \langle\theta_t, \mathrm{d}X_t\rangle = \sum_{i=1}^d \int_0^T \theta_t^i \mathrm{d}X_t^i$$

for $\theta \in \left(L^2_{\ell oc}(\mathrm{Tr}\langle\!\langle X, X\rangle\!\rangle)\right)^d$. Then by approximation procedure, we can define the stochastic integral of $\theta \in L^2_{\ell oc}(\langle\!\langle X, X\rangle\!\rangle)$.

Lemma 8.20. Assume $\theta \in L^2_{\ell oc}(\langle\!\langle X, X\rangle\!\rangle)$ and $\eta \in L^2_{\ell oc}(\langle\!\langle Y, Y\rangle\!\rangle)$. Then $\langle\theta_\bullet X, \eta_\bullet Y\rangle(\mathrm{d}t) = \theta_t^* \langle\!\langle X, Y\rangle\!\rangle(\mathrm{d}t)\eta_t$, where

$$\langle\!\langle X, Y\rangle\!\rangle = (\langle X^i, Y^j\rangle : 1 \leqslant i, j \leqslant d).$$

Proof When $\theta \in \left(L^2_{\ell oc}(\mathrm{Tr}\langle\!\langle X, X\rangle\!\rangle)\right)^d$ and $\eta \in \left(L^2_{\ell oc}(\mathrm{Tr}\langle\!\langle Y, Y\rangle\!\rangle)\right)$,

$$\langle\theta_\bullet X, \eta_\bullet Y\rangle = \sum_{i,j=1}^d \left\langle \int_0^\bullet \theta^i \mathrm{d}X^i, \int_0^\bullet \eta^j \mathrm{d}Y^j \right\rangle = \sum_{i,j=1}^d \theta^i \left\langle X^i, \int_0^\bullet \eta^j \mathrm{d}Y^j \right\rangle$$
$$= \sum_{i,j=1}^d \theta^i \eta_j \langle X^i, Y^j\rangle = \theta^* \langle\!\langle X, Y\rangle\!\rangle \eta.$$

The general case can be proved by using approximation.   □

Proposition 8.21. For $\theta \in L^2_{\ell oc}(\langle\!\langle X, X \rangle\!\rangle)$, $\theta_\bullet X$ is the unique $I \in \mathscr{M}^2_c$ such that

$$\left\langle I, \int_0^\bullet \eta \mathrm{d}X \right\rangle (\mathrm{d}t) = \theta_t^* \langle\!\langle X, X \rangle\!\rangle (\mathrm{d}t) \eta_t, \qquad \eta \in L^2_{\ell oc}(\langle\!\langle X, X \rangle\!\rangle).$$

Proof From Lemma 8.20, it follows that $I = \int_0^\bullet \theta \mathrm{d}X$ satisfies the above equation. The proof of uniqueness is much more difficult. We need to prove that every $I \in \mathscr{M}^2_c$ has the following expression

$$I = \sum_j \int_0^\bullet \gamma_j(t) \mathrm{d}X_t^j.$$

See Ikeda-Watanabe [39; §III.2].

Next, assume $\sigma : [0, \infty) \times \Omega \to \mathbb{R}^N \otimes \mathbb{R}^d$ is progressively measurable and

$$\mathbb{E}\left[\int_0^T \mathrm{Tr}\big(\sigma_t \langle\!\langle X, X \rangle\!\rangle (\mathrm{d}t) \sigma_t^* \big) \right] < \infty.$$

Then define $\sigma_\bullet X = \int_0^\bullet \sigma_t \mathrm{d}X(t) \in \big(\mathscr{M}^2_c \big)^N$ as

$$\left(\theta, \int_0^\bullet \sigma \mathrm{d}X \right)_{\mathbb{R}^N} = \int_0^\bullet \sigma^* \theta \mathrm{d}X, \qquad \text{a.s.,} \quad \theta \in \mathbb{R}^N.$$

Thus under the orthonormal basis $\{e_i\}$ of $\mathbb{R}^N$, $\sigma_\bullet X$ can be expressed as

$$\sum_{i=1}^N \left(\int_0^\bullet \sigma^* e_i \mathrm{d}X \right) e_i. \qquad \square$$

8.8 Stochastic Differential Equation (Multidimension)

Assume $\sigma : [0, \infty) \times \mathbb{R}^N \to \mathbb{R}^N \otimes \mathbb{R}^d$, $b : [0, \infty) \times \mathbb{R}^N \to \mathbb{R}^N$ and (B_t) is BM^d. We consider the following stochastic integral equation:

$$Z_T^x = x + \int_0^T \sigma\big(t, Z_t^x\big) \mathrm{d}B_t + \int_0^T b\big(t, Z_t^x\big) \mathrm{d}t, \qquad T \geqslant 0.$$

Formally, $Z^x : [0, \infty) \times \Omega \to \mathbb{R}^N$ can be written as the following stochastic differential equation:

$$\mathrm{d}Z_t^x = \sigma\big(t, Z_t^x\big) \mathrm{d}B_t + b\big(t, Z_t^x\big) \mathrm{d}t, \qquad Z_0^x = x.$$

Theorem 8.22. Suppose for each $T > 0$, there exists $C(T) < \infty$ such that

$$\sup_{0 \leqslant t \leqslant T} \|\sigma(t, 0)\|_{\text{H.S.}} \vee |b(t, 0)| \leqslant C(T),$$

$$\sup_{0 \leqslant t \leqslant T} \|\sigma(t, y') - \sigma(t, y)\|_{\text{H.S.}} \vee |b(t, y') - b(t, y)| \leqslant C(T)|y' - y|, \; y, y' \in \mathbb{R}^N.$$

(8.3)

(H.S. norm of matrix A is defined as $\|A\|^2_{H.S.} = \sum_{i,j} a_{ij}^2$.) Then the solution of the above equation exists and is unique.

Proof We prove the existence first by using the iteration method. Without loss of generality, assume $s = 0$, $x = 0$. Let $Z^{(0)} \equiv 0$,

$$Z_T^{(n)} = \int_0^T \sigma\big(t, Z_t^{(n-1)}\big)\mathrm{d}B_t + \int_0^T b\big(t, Z_t^{(n-1)}\big)\mathrm{d}t.$$

And let

$$\Delta_n(T) = \sup_{0 \leqslant t \leqslant T} |Z_t^{(n)} - Z_t^{(n-1)}|, \qquad T \geqslant 0.$$

Consider first the estimate of $\Delta_1(T)$. Let $\{e_i\}_{i=1}^N$ be the orthonormal basis of $\mathbb{R}^N$, and simply denote $\sigma = \sigma(\bullet, 0)$. Then

$$\mathbb{E}\left[\left|\int_0^T \sigma \mathrm{d}B\right|^2\right] = \sum_{i=1}^N \mathbb{E}\left[\left(\int_0^T \sigma^* e_i \mathrm{d}B_t\right)^2\right]$$

$$= \sum_{i=1}^N \mathbb{E}\left[\left(\sum_{j=1}^d \int_0^T (\sigma^* e_i)_j \mathrm{d}B_t^j\right)^2\right]$$

$$= \sum_{i=1}^N \mathbb{E}\left[\sum_{j,k} \int_0^T (\sigma^* e_i)_j \mathrm{d}B_t^j \int_0^T (\sigma^* e_i)_k \mathrm{d}B_t^k\right]$$

$$= \sum_{i=1}^N \sum_{j=1}^d \mathbb{E}\left[\left(\int_0^T (\sigma^* e_i)_j \mathrm{d}B_t^j\right)^2\right]$$

(by the property of independent increments of BM)

$$= \sum_{i=1}^N \sum_{j=1}^d \left[\int_0^T (\sigma^* e_i)_j^2 \mathrm{d}t\right] = \sum_{i=1}^N \sum_{j=1}^d \left[\int_0^T \sigma_{ij}^2 \mathrm{d}t\right]$$

$$= \int_0^T \|\sigma\|_{H.S.}^2 \mathrm{d}t.$$

Notice if we write $\int_0^t \sigma(u,0)\mathrm{d}B_u$ as $(Y^1, \cdots, Y^N)$ and $Y^j \in \mathcal{M}_c^2$, then

$$\left|\int_0^t \sigma \mathrm{d}B\right|^2 = \sum_{j=1}^N (Y^j)^2.$$

It is also a submartingale and we can apply Doob's inequality. Therefore,

$$\mathbb{E}[\Delta_1(T)^2] \leqslant 2\mathbb{E}\left[\sup_{0 \leqslant t \leqslant T}\left|\int_0^t \sigma(u,0)\mathrm{d}B_u\right|^2\right] + 2\mathbb{E}\left[\sup_{0 \leqslant t \leqslant T}\left|\int_0^t |b(u,0)|\mathrm{d}u\right|^2\right]$$

$$\leqslant 8\mathbb{E}\left[\left|\int_0^T \sigma(u,0)\mathrm{d}B_u\right|^2\right] + 2\mathbb{E}\left[T\int_0^T |b(u,0)|^2 \mathrm{d}u\right]$$

$$\leqslant 8\mathbb{E}\left[\int_0^T \|\sigma(t,0)\|_{H.S.}^2 \mathrm{d}t\right] + 2T^2 C(T)^2$$

$$\leqslant (8 + 2T)C(T)^2 T.$$

Similarly,

$$\mathbb{E}[\Delta_{n+1}(T)^2] \leqslant 8\mathbb{E}\left[\int_0^T \|\sigma(t, Z_t^{(n)}) - \sigma(t, Z_t^{(n-1)})\|_{H.S.}^2 dt\right]$$

$$+ 2T\mathbb{E}\left[\int_0^T |b(t, Z_t^{(n)}) - b(t, Z_t^{(n-1)})|^2 dt\right]$$

$$\leqslant (8 + 2T)C(T)^2\,\mathbb{E}\int_0^T |Z_t^{(n)} - Z_t^{(n-1)}|^2 dt$$

$$\leqslant (8 + 2T)C(T)^2 \int_0^T \mathbb{E}[\Delta_n(t)^2] dt.$$

By using the induction method, we have

$$\mathbb{E}[\Delta_n(T)^2] \leqslant K(T)^n/n!,$$

where $K(T) = (8 + 2T)TC(T)^2$. Then

$$\sup_{n \geqslant m} \mathbb{E}\left[\sup_{0 \leqslant t \leqslant T} |Z_t^{(n)} - Z_t^{(m)}|^2\right]$$

$$\leqslant \sup_{n \geqslant m} \mathbb{E}\left[\left(\sum_{\ell=m+1}^{n-1} \Delta_\ell(T)\right)^2\right] \leqslant \sum_{k,\ell=m+1}^{\infty} \mathbb{E}[\Delta_k(T)\Delta_\ell(T)]$$

$$\leqslant \sum_{k,\ell=m+1}^{\infty} (\mathbb{E}[\Delta_k(T)^2 \cdot \mathbb{E}[\Delta_\ell(T)^2])^{1/2}$$

$$\leqslant \left[\sum_{k=m+1}^{\infty} \left(\frac{K(T)^k}{k!}\right)^{1/2}\right]^2 \qquad \left(\text{by } \sum_k a_k b_k \leqslant \sum_k a_k \sum_k b_k\right)$$

$$\to 0, \qquad m \to \infty.$$

Therefore there exists a progressively measurable and a.s. continuous process Z, such that for each $T > 0$,

$$\mathbb{E}\left[\sup_{0 \leqslant t \leqslant T} |Z_t - Z_t^{(n)}|^2\right] \to 0, \qquad n \to \infty.$$

Now suppose there exist two solutions Z_t and Z_t', with initial values x and x', respectively. The previous proof gives us that $\Delta(T) = \sup_{0 \leqslant t \leqslant T} |Z_t - Z_t'|$ satisfies (by $(a + b + c)^2 \leqslant 3(a^2 + b^2 + c^2)$)

$$\mathbb{E}[\Delta(T)^2] \leqslant 3|x - x'|^2 + (12 + 3T)TC(T)^2 \int_0^T \mathbb{E}[\Delta(t)^2] dt.$$

Thus by Gronwall's lemma

$$\mathbb{E}[\Delta(T)^2] \leqslant 3|x - x'|^2 \exp\left[(12 + 3T)TC(T)^2\right].$$

In particular, when $x = x'$ we obtain the uniqueness. $\quad\square$

Remark 8.23. In [22], condition (8.3) is weakened to the following non-Lipschitz condition, which can still imply the uniqueness and existence: for any $y', y : |y' - y| < 1$ satisfying

$$\sup_{0 \leqslant t \leqslant T} \|\sigma(t, 0)\|_{\text{H.S.}} \vee |b(t, 0)| \leqslant C(T),$$

$$\sup_{0 \leqslant t \leqslant T} \|\sigma(t, y') - \sigma(t, y)\|_{\text{H.S.}}^2 \leqslant C(T)|y' - y|^2 r(|y' - y|^2),$$

$$\sup_{0 \leqslant t \leqslant T} |b(t, y') - b(t, y)| \leqslant C(T)|y' - y|r(|y' - y|^2),$$

where $r \in C^1((0, 1), \mathbb{R}_+)$ satisfies

$$\lim_{s \to 0} \frac{sr'(s)}{r(s)} = 0 \quad \text{and} \quad \int_0^a \frac{ds}{sr(s)} = \infty \quad \text{for arbitrary} \quad a > 0.$$

Now suppose $\sigma(t, x) = \sigma(x)$, $b(t, x) = b(x)$ is independent of t. Consider equation

$$X_T = X_0 + \int_0^T \sigma(X_t)dB_t + \int_0^T b(X_t)dt,$$

where (B_t) is BM^d. Let

$$\bar{X}_T = X_T - \int_0^T b(X_t)dt \in \mathcal{M}_c^2.$$

Then

$$\langle\langle \bar{X}, \bar{X} \rangle\rangle = \left\langle\!\!\left\langle \int_0^\bullet \sigma(X_t)dB_t, \int_0^\bullet \sigma(X_t)dB_t \right\rangle\!\!\right\rangle = \sigma\sigma^*(X_t) =: a(X_t).$$

Apply Itô's formula to

$$F(\bar{X}_T, Y_T) = f(\bar{X}_T + Y_T), \qquad Y_T = \int_0^T b(X_t)dt, \quad f \in C_b^2(\mathbb{R}^N)$$

to derive

$$f(X_T) - f(X_0) - \int_0^T Lf(X_t)dt = \int_0^T \nabla f(X_t)d\bar{X}_t \in \mathcal{M}_c^2,$$

where

$$L = \frac{1}{2} \sum_{i,j=1}^N a_{ij}(x)\partial_{ij}^x + \sum_{i=1}^N b_i(x)\partial_i^x.$$

Thus for each $f \in C_b^2(\mathbb{R}^N)$, $\left(f(X_t) - \int_0^t Lf(X_u)du, \mathscr{F}_t, \mathbb{P} \right)$ is a martingale. This establishes the relation between diffusion process (X_t) and the second order differential operator L. We will come back to this topic in the next section.

8.9 Feynman-Kac Formula, Random Change of Time, and Girsanov's Theorem

Those basic mathematical tools in the title above are three magic weapons in stochastic analysis, which have extensive applications. Here, we restrict ourselves on the basic part, to explain the main ideas inside. For more applications, see [31; 50]. In the last part of this section, we apply them to the convex geometry, and introduce a probabilistic proof of the famous Brunn-Minkowski's inequality. This presents an excellent example for applying probability theory to geometry and analysis.

At the end of previous section, we use the stochastic differential equation to construct a diffusion process $(X_t)_{t \geqslant 0}$ corresponding to the operator

$$L = \frac{1}{2} \sum_{i,j=1}^{N} a_{ij}(x) \partial_{ij}^x + \sum_{i=1}^{N} b_i(x) \partial_i^x,$$

where $a(x) = (\sigma \sigma^*)(x)$. The following result presents a stochastic expression of the solution to a parabolic equation.

Theorem 8.24 (Feynman-Kac formula). Assume $V \in {}_b\mathscr{B}(\mathbb{R}^N)$ and $\phi \in C_b(\mathbb{R}^N)$. Suppose $f \in C_b^{1,2}([0,\infty) \times \mathbb{R}^N)$ satisfies equation

$$\begin{cases} \dfrac{\partial f}{\partial t} = Lf + Vf, \\ f(0, \cdot) = \phi. \end{cases}$$

Then f has the following probabilistic representation:

$$f(t,x) = \mathbb{E}_x \left\{ \phi(X_t) \exp\left[\int_0^t V(X_s) \mathrm{d}s \right] \right\}, \quad t \geqslant 0, \ x \in \mathbb{R}^N.$$

Proof Fix $t_0 > 0$. Apply Itô's formula to function

$$(t, z_1, z_2) \to f(t_0 - t, z_1) e^{z_2}, \quad t \leqslant t_0, \ z_1 \in \mathbb{R}^N, \ z_2 \in \mathbb{R}$$

to derive

$$f(t_0 - t, X_t) \exp\left[\int_0^t V(X_s) \mathrm{d}s \right] - f(t_0, x)$$

$$= \int_0^t \langle \nabla^x f, \mathrm{d}\bar{X}_s \rangle \exp\left[\int_0^s V(X_u) \mathrm{d}u \right] \mathrm{d}s$$

$$+ \int_0^t f(t_0 - s, X_s) \exp\left[\int_0^s V(X_u) \mathrm{d}u \right] V(X_s) \mathrm{d}s \qquad (8.4)$$

$$+ \int_0^t \left(-\frac{\partial}{\partial s} f + Lf \right)(s, X_s) \exp\left[\int_0^s V(X_u) \mathrm{d}u \right] \mathrm{d}s$$

$$= \int_0^t \langle \nabla^x f, \mathrm{d}\bar{X}_s \rangle \exp\left[\int_0^s V(X_u) \mathrm{d}u \right] \mathrm{d}s \in \mathscr{M}_c^2,$$

where $(\bar{X}_t)$ was defined in the previous section:

$$\bar{X}_t = X_t - \int_0^t b(X_s)\mathrm{d}s \in \mathcal{M}_c^2.$$

Making expectation in both sides of (8.4), we have

$$f(t_0, x) = \mathbb{E}_x \left\{ f(t_0 - t, X_t) \exp\left[\int_0^t V(X_s)\mathrm{d}s \right] \right\}.$$

Letting $t \uparrow t_0$, we obtain the required conclusion with the initial value $f(0, \cdot) = \phi$.   □

A much simpler case is the Dirichlet boundary value problem. Suppose D is a bounded open subset of $\mathbb{R}^N$. Let

$$\tau = \tau_D = \inf\{t \geqslant 0 : B_t \notin D\}.$$

Theorem 8.25. Suppose $\phi \in {}_b\mathcal{B}(\partial D)$ and $f \in C_b^2(\mathbb{R}^N)$ satisfy

$$\begin{cases} \Delta f = 0 & \text{inside of } D, \\ f|_{\partial D} = \phi. \end{cases}$$

Then f can be expressed as

$$f(x) = \int_{\partial D} \phi(y)\mathbb{P}_x^{B(\tau)}(\mathrm{d}y),$$

where $\mathbb{P}_x^{B(\tau)}$ is the distribution on boundary ∂D at the first hitting ∂D of BM^N, starting at x.

Proof Since

$$f(B_t) - \frac{1}{2}\int_0^t \Delta f(B_s)\mathrm{d}s$$

is a martingale, by Doob's stopping theorem

$$f(B_\tau) - \frac{1}{2}\int_0^{t \wedge \tau} \Delta f(B_s)\mathrm{d}s$$

is also a martingale. Thus

$$f(x) = \mathbb{E}_x\left[f(B_\tau)\right] = \mathbb{E}_x\left[\phi(B_\tau)\right] = \int_{\partial D} \phi(y)\mathbb{P}_x^{B(\tau)}(\mathrm{d}y).   □$$

The above two results present the relation between the diffusion process and partial differential equation. Indeed, both are the descriptions of the physical systems, the former is microscopic and the latter is macroscopic. However, the microscopic description is much finer and more difficult. Its obvious advantage is: for example, in the above result the boundary ∂D of domain D can be irregular, it can be a much rougher boundary (for example, fractal set). Moreover the conditions on coefficients of a, b, V can be relaxed such that the smoothness is weaker than the usual condition used in analysis. This shows why the probability theory has become an important research tool in analysis.

Now, we turn to the second topic in this section—**random change of time**. Indeed, we have used this technique when defining the stochastic integral in Section 4. Here we introduce another example.

As we know, the process $(f(B_t))_{t \geqslant 0}$ is not a BM for the nonlinear and smooth real function f. The following results is rather wonderful which shows that the complex space has much more "inflexibility" than the real space.

We call $B_t = B_t^1 + \sqrt{-1} B_t^2$ is a (standard) **complex BM**, if (B_t^1) and (B_t^2) are independent BM. Recall that the complex function $f = u + \sqrt{-1}v$ is called analytic function, if it can be expanded to an absolutely convergent power series. Equivalently, the real part u and imaginary part v of f satisfy the following Cauchy-Riemann condition:

$$u_x = v_y, \quad u_y = -v_x.$$

Proposition 8.26. Assume $(B_t)_{t \geqslant 0}$ is a standard complex BM and f is a analytic function. Let σ_t be the unique solution of equation

$$\int_0^{\sigma_t} |f'(B_s)|^2 ds = t, \qquad t > 0.$$

Then $(f(B_{\sigma_t}))_{t \geqslant 0}$ is also a standard complex BM.

Proof Note that both the real part u and imaginary part v of f are harmonic functions: $\Delta u = 0$, $\Delta v = 0$. It follows from the Itô's formula that

$$du(B_t) = u_x(B_t)dB_t^1 + u_y(B_t)dB_t^2,$$
$$dv(B_t) = v_x(B_t)dB_t^1 + v_y(B_t)dB_t^2,$$
$$= -u_y(B_t)dB_t^1 + u_x(B_t)dB_t^2.$$

Thus,

$$\langle u(B) \rangle_t = \langle v(B) \rangle_t = \int_0^t (u_x^2 + u_y^2)(B_s)ds = \int_0^t |f'(B_s)|^2 ds,$$

$$\langle u(B), v(B) \rangle = 0,$$

which implies the required assertion. $\qquad \square$

More generally, if $\phi \in \mathscr{B}(\mathbb{R}^N)$ satisfies

$$0 < c_1 \leqslant \phi \leqslant c_2 < \infty,$$

then we can define **random change of time** τ_ϕ, which is the unique solution of equation

$$\int_0^{\tau_\phi(t)} \frac{ds}{\phi(X_s)} = t, \qquad t > 0.$$

When coefficient a is uniformly positive definitive and bounded, and $b = 0$, the diffusion process (X_t) can be transformed to Brownian motion via random time change $\tau_\phi : \phi = a^{-1}$. For more details, see [63; §6.5].

For general diffusion process $(X_t)_{t \geqslant 0}$, if the coefficient b in the first order of its operator L is not null, it can be changed to null by a measure transform. This is the third topic we are going to study—**Girsanov's transform**. Here we deal with a special case.

Assume $T \leqslant \infty$ and let

$$\Omega_T = \{\omega : \omega_0 = 0, \ \omega_t \ \text{continuous on} \ [0, T]\}, \quad \mathscr{F}_t^0 = \sigma(\omega_s : s \leqslant t).$$

Since the trajectories are a.s. continuous, we can regard the standard BM^N as a stochastic process $(B_t = \omega_t)_{t \in [0,T]}$, defined on $(\Omega_T, (\mathscr{F}_t^0)_{t \in [0,T]})$ with the Wiener measure $\mathbb{P}$. The advantage for restricting on $T < \infty$ is that Ω_T is a Banach space under the supremum norm.

In the following we fix $T < \infty$. Suppose that we are given a stochastic process $(u_t)_{t \in [0,T]}$ that is adapted with respect to $(\mathscr{F}_t^0)_{t \in [0,T]}$ and valued in $\mathbb{R}^N$, satisfying $\int_0^T |u_t|^2 dt < \infty$, a.s. and $\mathbb{E}_\mathbb{P} Z_T = 1$, where

$$Z_t = \exp\left[-\int_0^t u_s dB_s - \frac{1}{2} \int_0^t |u_s|^2 ds \right], \quad t \in [0, T].$$

On $\mathscr{F}_T$, we can define the following measure transform (**Girsanov's transform**)

$$d\mathbb{Q} = Z_T d\mathbb{P}.$$

Let

$$\tilde{B}_t = B_t + \int_0^t u_s ds, \quad t \in [0, T].$$

Theorem 8.27 (Girsanov transform theorem). On $(\Omega_T, \mathscr{F}_T^0, \mathbb{Q})$, $(\tilde{B}_t)_{t \in [0,T]}$ is a BM^N. Namely, any finite-dimensional distribution of $(\tilde{B}_t)$ before T under $\mathbb{Q}$ coincides with that of (B_t) under $\mathbb{P}$ (Wiener measure).

Proof We prove only the case of $N = 1$; the proof of the multi-dimensional case is similar and left as an exercise. Let

$$\tau_k = \inf\left\{ t : \int_0^t u_s^2 ds \geqslant k \right\}.$$

Then it follows from Itô formula that $(Z_{t \wedge \tau_k})_{t \in [0,T]}$ is a positive continuous martingale. Thus $(Z_t)_{t \in [0,T]}$ is a continuous local martingale. By $\mathbb{E}_P Z_T = 1$, it is indeed a martingale.

To prove the property of finite dimensions, since Brownian motion has independent increments with normal distributions, we need only prove that the function F having the following form

$$F(x_\bullet) = \exp\left[\sqrt{-1} \sum_{j=0}^{m-1} \lambda_j (x_{t_{j+1}} - x_{t_j}) \right], \quad \lambda_j \in \mathbb{R},\ 0 = t_0 < \cdots < t_m = T$$

satisfies $\mathbb{E}_P\left[F(\tilde{B}) Z_T \right] = \mathbb{E}_P F(B)$. For this, we first prove that

$$\mathbb{E}_P\left[\exp\left[\sqrt{-1}\,\lambda(\tilde{B}_t - \tilde{B}_s) \right] h Z_T \right] = \mathbb{E}_P\left[h Z_T \right] \exp\left[-\frac{\lambda^2}{2}(t - s) \right]$$

for every pair of $s < t\,(\leqslant T)$ and $h \in \mathscr{F}_s^0$. This can be deduced by applying Itô's formula to

$$\exp\left[\sqrt{-1}\,\lambda(B_t - B_s + \int_s^t u_r dr) - \int_s^t u_r dB_r - \frac{1}{2}\int_s^t u_r^2 dr \right]$$

to derive

$$\mathbb{E}_P\left\{ \exp\left[\sqrt{-1}\,\lambda(\tilde{B}_t - \tilde{B}_s) \right] h Z_T \right\}$$

$$= \mathbb{E}_P\left\{ \mathbb{E}_P\left[\exp\left[\sqrt{-1}\,\lambda(\tilde{B}_t - \tilde{B}_s) - \int_s^t u_r dB_r - \frac{1}{2}\int_s^t u_r^2 dr \right] \Big| \mathscr{F}_s^0 \right] h Z_T \right\}$$

$$= \mathbb{E}_P\left[h Z_T \right] - \frac{\lambda^2}{2}\int_s^t \mathbb{E}_P\left[\exp\left[\sqrt{-1}\,\lambda(\tilde{B}_t - \tilde{B}_s) \right] h Z_T \right] d\sigma.$$

By regarding the left-hand side as a function of t, it coincides with the integrand in the second term of the right-hand side. Then this becomes a simple integral equation, whose solution is just what we want. Here in the last step, we have used the martingale property of $\{Z_t\}$: for each $\xi \in \mathscr{F}_t^0$,

$$\mathbb{E}_P[\xi Z_T] = \mathbb{E}_P[\xi \mathbb{E}_P[Z_T | \mathscr{F}_0^0]] = \mathbb{E}_P[\xi Z_t].$$

For general F, we apply the above result to

$$h = \exp\left[-\sqrt{-1} \sum_{j=0}^{m-2} \lambda_j \left(\tilde{B}_{t_{j+1}} - \tilde{B}_{t_j} \right) \right]$$

to derive

$$\mathbb{E}_P\left\{ \exp\left[-\sqrt{-1} \sum_{j=0}^{m-1} \lambda_j \left(\tilde{B}_{t_{j+1}} - \tilde{B}_{t_j} \right) \right] Z_T \right\}$$

$$= \mathbb{E}_P\left\{ e^{-\sqrt{-1}\lambda_{m-1}\left(\tilde{B}_{t_m} - \tilde{B}_{t_{m-1}} \right)} h Z_T \right\}$$

$$= \mathbb{E}_P\left[h Z_t \right] \exp\left[-\frac{1}{2}\lambda_{m-1}^2(t_m - t_{m-1}) \right].$$

Inductively, we obtain

$$\mathbb{E}_{\mathbb{P}}\left[F(\tilde{B})Z_T\right] = \mathbb{E}(Z_T)\exp\left[-\frac{1}{2}\sum_{j=0}^{m-1}\lambda_j^2(t_{j+1}-t_j)\right]$$

$$= \exp\left[-\frac{1}{2}\sum_{j=0}^{m-1}\lambda_j^2(t_{j+1}-t_j)\right]$$

$$= \mathbb{E}_{\mathbb{P}}F(B). \quad \square$$

Now we apply the previous results to a variational description for the solution of a kind of simple diffusion equation, which prepares us for the next topic. Assume σ is a positive constant and $c, f \in C_b(\mathbb{R}^n)$ with $\inf_x f > 0$. Consider the Cauchy problem

$$\begin{cases} \dfrac{\partial v}{\partial t} = \dfrac{\sigma^2}{2}\Delta v - \dfrac{1}{\sigma^2}c(x)v, \ t > 0, x \in \mathbb{R}^n, \\ v(0,x) = f(x), \ x \in \mathbb{R}^n. \end{cases}$$

By the transforms $V = -\sigma^2 \log v$ and $F = -\sigma^2 \log f$, this becomes the Cauchy problem for Hamilton-Jacobi-Bellman equation

$$\begin{cases} \dfrac{\partial V}{\partial t} + \dfrac{1}{2}|\nabla V|^2 - c(x) = \dfrac{\sigma^2}{2}\Delta V, \quad t > 0, x \in \mathbb{R}^n, \\ V(0,x) = F(x), \quad x \in \mathbb{R}^n. \end{cases}$$

Let $B_t^{x,\sigma} = x + \sigma B_t$, $t \geq 0$. Then the Feynman-Kac formula gives

$$v(t,x) = v_{\sigma,c}^F(t,x) = \mathbb{E}_{\mathbb{P}}\left\{\exp\left[-\frac{1}{\sigma^2}F(B_t^{x,\sigma}) + \int_0^t c(B_s^{x,\sigma})\,\mathrm{d}s\right]\right\},$$

where $\mathbb{P}$ is the Wiener measure.

The following result describes further a variational formula for $v_{\sigma,c}^F$.

Proposition 8.28. Denote by $\mathscr{U}(T)$ the total of all bounded and progressively measurable processes $(u_t)_{t\in[0,T]}$. Assume $c, F \in C_b^\infty(\mathbb{R}^n)$. For each $u \in \mathscr{U}(T)$, let

$$h(t) = h_u(t) = \int_0^t u_s\,\mathrm{d}s, \quad t \in [0,T],$$

$$J_{\sigma,c}^F(t,x,u) = \frac{1}{\sigma^2}\mathbb{E}_{\mathbb{P}}\left[F(B_t^{x,\sigma}+h_u(t)) + \int_0^t\left(c(B_s^{x,\sigma}+h_u(s)) + \frac{1}{2}|u_s|^2\right)\mathrm{d}s\right].$$

Then

$$v_{\sigma,c}^F(T,x) = \exp\left[-\inf_{u\in\mathscr{U}(T)} J_{\sigma,c}^F(T,x,u)\right], \quad T > 0, x \in \mathbb{R}^n.$$

Proof Let

$$X_t = X_t^u = B_t^{x,\sigma} + h_u(t), \qquad t \in [0, T],$$

$$Y_t^u = F(X_t) + \int_0^t \left(c(X_s) + \frac{1}{2}|u_s|^2 \right) ds + \sigma \int_0^t u_s dB_s, \qquad t \in [0, T].$$

For $d\mathbb{Q} = Z_T d\mathbb{P}$, where

$$Z_T = \exp\left[-\frac{1}{2\sigma^2} \int_0^T |u_s|^2 ds - \frac{1}{\sigma} \int_0^t u_s dB_s \right],$$

and nonnegative $\mathscr{F}_T^0$-measurable function ϕ, by Girsanov's theorem, we have

$$\int_{\Omega_T} \phi\left(\omega + \frac{h}{\sigma} \right) \mathbb{Q}(d\omega) = \int_{\Omega_T} \phi(\omega)\mathbb{P}(d\omega).$$

Notice $B_t^{x,\sigma}$ becomes $x + \sigma(B_t + h(t)/\sigma) = B_t^{x,\sigma} + h(t) = X_t$ via the transform $B_t = \omega_t \to \omega_t + h(t)/\sigma$. We have

$$v(T, x) = \mathbb{E}_\mathbb{P} \exp\left[-\frac{1}{\sigma^2} \left\{ F\left(B_T^{x,\sigma}\right) + \int_0^T c\left(B_t^{x,\sigma}\right) dt \right\} \right]$$

(Feynman-Kac formula)

$$= \mathbb{E}_\mathbb{Q} \exp\left[-\frac{1}{\sigma^2} \left\{ F(X_T) + \int_0^T c(X_t) dt \right\} \right]$$

$$= \mathbb{E}_\mathbb{P} \exp\left[-\frac{1}{\sigma^2} Y_T^u \right]. \qquad \text{(by definition of } \mathbb{Q})$$

Thus it follows from Jensen's inequality that

$$\log v(T, x) \geqslant -\frac{1}{\sigma^2} \mathbb{E}_\mathbb{P} Y_T^u.$$

For fixed u, if Y_u is constant almost everywhere, then this inequality becomes equality. In other words, we can choose a suitable function η in Girsanov's theorem such that Y^η is constant almost everywhere, which is a way for optimizing $v(T, x)$. To do so, let

$$U(t, x) = -\nabla V(T - t, x), \qquad t \in [0, T].$$

Then the boundedness and smoothness of c and F guarantee that $U(t, \cdot)$ is Lipschitz. Therefore, equation

$$dX_t = U(t, X_t)dt + \sigma dB_t, \qquad t \in [0, T], X_0 = x$$

has a unique solution. By taking $\eta_t = U(t, X_t)$ for $t \in [0, T]$, we obtain

$$X_t = X_t^\eta = x + \sigma B_t + h_\eta(t) = B_t^{x,\sigma} + h_\eta(t).$$

And this coincides with X_t as above.

Consider the process

$$\xi_t = V(T - t, X_t) + \int_0^t \left(c\left(X_s \right) + \frac{1}{2}|\eta_s|^2 \right) ds + \sigma \int_0^t \eta_s dB_s.$$

It follows from Itô's formula that

$$d\xi_t = -\frac{\partial}{\partial t}V(T - t, X_t)dt + \nabla V(T - t, X_t) \cdot \left(\eta_t dt + \sigma dB_t \right)$$

$$+ \frac{\sigma^2}{2}\Delta V(T - t, X_t)dt + \left(c(X_t) + \frac{1}{2}|\eta_t|^2 \right) dt + \sigma \eta_t dB_t$$

$$= -\frac{\partial}{\partial t}V(T - t, X_t)dt - \eta_t \cdot \left(\eta_t dt + \sigma dB_t \right)$$

$$+ \frac{\sigma^2}{2}\Delta V(T - t, X_t)dt + \left(c(X(t)) + \frac{1}{2}|\eta_t|^2 \right) dt + \sigma \eta_t dB_t$$

$$= \left[-\frac{\partial}{\partial t}V(T - t, X_t) - \frac{1}{2}|\nabla V(T - t, X_t)|^2 + \frac{\sigma^2}{2}\Delta V(T - t, X_t) + c(X_t) \right] dt$$

$$= 0, \qquad a.s. \quad t \leqslant T.$$

The final step comes from the fact that $V(T, x)$ satisfies Hamilton-Jacobi-Bellman equation. Thus, $\xi_T = Y_T^\eta$ is constant almost everywhere. From this and $\mathbb{E}_\mathbb{P} \int_0^t u_s dB_s = 0$, we obtain the assertion as required. $\qquad \square$

Now let us turn to the final topic, an important inequality in convex geometry. In the following, assume $\theta = (\theta_0, \theta_1) \in \mathbb{R}^2$ is a positive vector. Let $x_\theta = \theta_0 x_0 + \theta_1 x_1$ for $x_0, x_1 \in \mathbb{R}^n$. For given subsets A_0 and A_1 of $\mathbb{R}^n$, denote $A_\theta = \{x_\theta : x_0 \in A_0, x_1 \in A_1\}$, or write it simply as $A_\theta = \theta_0 A_0 + \theta_1 A_1$. Let $V_n(A)$ be the volume (Lebesgue measure) of $A \in \mathscr{B}(\mathbb{R}^n)$. Then the famous **Brunn-Minkowski's inequality** says that

$$V_n(A_\theta) \geqslant V_n(A_0)^{\theta_0} V_n(A_1)^{\theta_1}, \qquad \theta_0 + \theta_1 = 1, \ A_0, A_1 \in \mathscr{B}(\mathbb{R}^n). \quad (8.5)$$

This inequality has several equivalent forms

$$V_n(A_\theta)^{1/n} \geqslant \theta_0 V_n(A_0)^{1/n} + \theta_1 V_n(A_1)^{1/n}, \quad (8.6)$$

$$V_n(A_\theta)^{1/n} \geqslant \min\{V_n(A_0), V_n(A_1)\}, \qquad \theta_0 + \theta_1 = 1, \ A_0, A_1 \in \mathscr{B}(\mathbb{R}^n).$$

The equivalence is not difficult to prove. For example, by using arithmetic-geometric mean inequality, (8.6) implies (8.5). Conversely, in (8.5) take $A_0 = V_n(A_0')^{-1/n} A_0'$ and $A_1 = V_n(A_1')^{-1/n} A_1'$, such that the right-hand side of (8.5) equals 1. Next, take

$$\theta_0 = \frac{V_n(A_0')^{1/n}}{V_n(A_0')^{1/n} + V_n(A_1')^{1/n}}.$$

Then the left-hand side of (8.5) becomes

$$V_n\left(\frac{A_0' + A_1'}{V_n(A_0')^{1/n} + V_n(A_1')^{1/n}}\right) = \frac{V_n(A_0' + A_1')}{(V_n(A_0')^{1/n} + V_n(A_1')^{1/n})^n}.$$

This implies

$$V_n(A_0' + A_1')^{1/n} \geqslant V_n(A_0')^{1/n} + V_n(A_1')^{1/n}.$$

Finally by taking $A_0' = \theta_0 A_0$ and $A_1' = \theta_1 A_1$, then we come back to (8.6). For more information and recent processes for this inequality, see [26]. This inequality is a basic result in convex geometry, the reason for this is that, when it was discovered, both A_0 and A_1 are assumed to be convex, then A_θ is a convex combination of A_0 and A_1.

In the following, we will use a functional inequality. Assume $D_i (i = 0, 1)$ are subsets of $\mathbb{R}^n$. Let $\sigma_\theta = \theta_0 \sigma_0 + \theta_1 \sigma_1$ for $\sigma_0, \sigma_1 > 0$. We consider such functions $\phi_j : D_j \to \mathbb{R}$ $(j = 0, 1, \theta)$ satisfying the following relation

$$\frac{1}{\sigma_\theta} \phi_\theta(x_\theta) \leqslant \frac{\theta_0}{\sigma_0} \phi_0(x_0) + \frac{\theta_1}{\sigma_1} \phi_1(x_1), \quad x_0 \in D_0, \, x_1 \in D_1. \quad (8.7)$$

The following three examples are those kinds of functions.

(1) $\sigma_0 = \sigma_1$, $\theta_0 + \theta_1 = 1$,

$$\phi_\theta(x_\theta) \leqslant \theta_0 \phi_0(x_0) + \theta_1 \phi_1(x_1), \quad x_0 \in D_0, \, x_1 \in D_1.$$

(2) $\phi_j = \psi_j^2$, $\psi_j \geqslant 0$, $j = 0, 1, \theta$ and

$$\psi_\theta(x_\theta) \leqslant \theta_0 \psi_0(x_0) + \theta_1 \psi_1(x_1), \quad x_0 \in D_0, \, x_1 \in D_1.$$

(3) $\phi_j = \sigma_j^4/\psi_j^2$, $j = 0, 1, \theta$, ψ_j satisfies

$$\psi_\theta(x_\theta) \geqslant \theta_0 \psi_0(x_0) + \theta_1 \psi_1(x_1), \quad x_0 \in D_0, \, x_1 \in D_1.$$

To check the last two cases, consider function

$$\gamma_\alpha(\lambda, \sigma) = \lambda^{\alpha+1}/\sigma^\alpha, \quad \lambda \geqslant 0, \, \sigma > 0.$$

It is first order homogeneous: $\gamma_\alpha(c\lambda, c\sigma) = c\gamma_\alpha(\lambda, \sigma)$ for every $c > 0$. When $\alpha = 1$, it is convex and

$$\text{Hess}(\gamma_\alpha) = \alpha(\alpha + 1)\frac{\lambda^{\alpha-1}}{\sigma^\alpha}\begin{pmatrix} 1 & -\lambda/\sigma \\ -\lambda/\sigma & \lambda^2/\sigma^2 \end{pmatrix} = 0.$$

The next result is the main theorem in [3].

Theorem 8.29. Assume $\sigma_0, \sigma_1 > 0$, c_j and F_j $(j = 0, 1, \theta)$ are bounded continuous functions, and (8.7) holds for every $x_0, x_1 \in \mathbb{R}^n$. Then for every $x_0, x_1 \in \mathbb{R}^n$ and $A_0, A_1 \in \mathscr{B}(\mathbb{R}^n)$, we have

$$v^{A_\theta, F_\theta}_{\sigma_\theta, c_\theta}(T, x_\theta) \geqslant \left\{v^{A_0, F_0}_{\sigma_0, c_0}(T, x_0)\right\}^{\theta_0 \sigma_0/\sigma_\theta} \left\{v^{A_1, F_1}_{\sigma_1, c_1}(T, x_1)\right\}^{\theta_1 \sigma_1/\sigma_\theta},$$

where $\sigma_\theta = \theta_0 \sigma_0 + \theta_1 \sigma_1$, and

$$v^{A, F}_{\sigma, c}(T, x) = \mathbb{E}^{\mathbb{P}}\left[I_A(B_T^{x, \sigma}) \exp\left[-\frac{1}{\sigma^2}\left\{F(B_T^{x, \sigma}) + \int_0^T c(B_s^{x, \sigma}) \, ds\right\}\right]\right].$$

As a simple corollary of Theorem 8.29, by assuming $\theta_0 + \theta_1 = 1, \sigma_0 = \sigma_1 = \sigma$ and taking $c_0 = c_1 = c_\theta = 0, F_0 = F_1 = F_\theta = 0$ (that is $f_j \equiv 1$), $x_0 = x_1 = 0$, we have

$$\mathbb{E}^{\mathbb{P}}\left[I_{A_\theta}\left(B_T^{0,\sigma}\right)\right] \geqslant \left\{\mathbb{E}^{\mathbb{P}}\left[I_{A_0}\left(B_T^{0,\sigma}\right)\right]\right\}^{\theta_0}\left\{\mathbb{E}^{\mathbb{P}}\left[I_{A_1}\left(B_T^{0,\sigma}\right)\right]\right\}^{\theta_1}.$$

When $T = 1$, we have the following "Gaussian" type inequality:

$$\int_{A_\theta} e^{-\frac{|x|^2}{2\sigma^2}}\,\mathrm{d}x \geqslant \left\{\int_{A_0} e^{-\frac{|x|^2}{2\sigma^2}}\,\mathrm{d}x\right\}^{\theta_0}\left\{\int_{A_1} e^{-\frac{|x|^2}{2\sigma^2}}\,\mathrm{d}x\right\}^{\theta_1}.$$

Letting $\sigma \to \infty$, we obtain the original Brunn-Minkowski's inequality (8.5). Note that the latter is the limit form of the former "Gaussian" type inequality.

Let us make a little change. Assume $f_j\ (j = 0, 1, \theta)$ satisfies

$$f_\theta(x_\theta) \geqslant f_0(x_0)^{\theta_0} f_1(x_1)^{\theta_1}, \quad x_0, x_1 \in \mathbb{R}^n.$$

Without loss of generality, assume f_j is bounded and positive. It follows from Theorem 8.29 that

$$\int_{A_\theta} f_\theta(x)e^{-\frac{|x|^2}{2\sigma^2}}\,\mathrm{d}x \geqslant \left\{\int_{A_0} f_0(x)e^{-\frac{|x|^2}{2\sigma^2}}\,\mathrm{d}x\right\}^{\theta_0}\left\{\int_{A_1} f_1(x)e^{-\frac{|x|^2}{2\sigma^2}}\,\mathrm{d}x\right\}^{\theta_1}.$$

By letting $\sigma \to \infty$, we get the Prékopa-Leindler's inequality:

$$\int_{A_\theta} f_\theta(x)\,\mathrm{d}x \geqslant \left\{\int_{A_0} f_0(x)\,\mathrm{d}x\right\}^{\theta_0}\left\{\int_{A_1} f_1(x)\,\mathrm{d}x\right\}^{\theta_1}.$$

Formally, this is a reverse Hölder's inequality, which is again equivalent to the Brunn-Minkowski's inequality. Thus it can be regarded as a function form of the Brunn-Minkowski's inequality.

Proof of Theorem 8.29   a) First prove the special case of $v_{\sigma,c}^F := v_{\sigma,c}^{\mathbb{R}^n,F}$. By using approximation by smooth functions, without loss of generality assume $c, F \in C_b^\infty(\mathbb{R}^n)$. Let $u_\theta = \theta_0 u_0 + \theta_1 u_1$ for $u_0, u_1 \in \mathscr{U}(T)$, and $X_t^{x,u,\sigma} = B_t^{x,\sigma} + h_u(t)$. Then

$$B_t^{x_\theta,\sigma_\theta} = x_\theta + \sigma_\theta B(t) = \theta_0 B_t^{x_0,\sigma_0} + \theta_1 B_t^{x_1,\sigma_1},$$

and

$$\begin{aligned}
X^{x_\theta,u_\theta,\sigma_\theta} &= B^{x_\theta,\sigma_\theta} + h_{u_\theta} = \theta_0\left(B^{x_0,\sigma_0} + h_{u_0}\right) + \theta_1\left(B^{x_1,\sigma_1} + h_{u_1}\right)\\
&= \theta_0 X^{x_0,u_0,\sigma_0} + \theta_1 X^{x_1,u_1,\sigma_1}.
\end{aligned}$$

Since c_j and F_j $(j = 0, 1, \theta)$ satisfy condition (8.7), by using the second example below (8.7) $(\phi_j = \psi_j^2)$ and taking $\psi_j = |u_j|$ $(j = 0, 1, \theta)$, we have

$$\frac{1}{\sigma_\theta} \left[F_\theta \left(X_T^{x_\theta, \sigma_\theta} \right) + \int_0^T \left(c_\theta \left(X_t^{x_\theta, \sigma_\theta} \right) + \frac{1}{2} |u_\theta(t)|^2 \right) dt \right]$$

$$\leqslant \frac{\theta_0}{\sigma_0} \left[F_0 \left(X_T^{x_0, \sigma_0} \right) + \int_0^T \left(c_0 \left(X_t^{x_0, \sigma_0} \right) + \frac{1}{2} |u_0(t)|^2 \right) dt \right]$$

$$+ \frac{\theta_1}{\sigma_1} \left[F_1 \left(X_T^{x_1, \sigma_1} \right) + \int_0^T \left(c_1 \left(X_t^{x_1, \sigma_1} \right) + \frac{1}{2} |u_1(t)|^2 \right) dt \right].$$

Take expectation $\mathbb{E}^{\mathbb{P}}$ in both sides to derive

$$\sigma_\theta J_{\sigma_\theta, c_\theta}^{F_\theta} (T, x_\theta, u_\theta) \leqslant \theta_0 \sigma_0 J_{\sigma_0, c_0}^{F_0} (T, x_0, u_0) + \theta_1 \sigma_1 J_{\sigma_1, c_1}^{F_1} (T, x_1, u_1).$$

Notice

$$-\sigma \log v_{\sigma, c}^F(T, x) = \inf_{u \in \mathscr{U}(T)} \sigma J(t, x, u) \quad \text{(Proposition 8.28)}.$$

The proves the assertion as required.

b) Next, we prove the general case. By the inner regularity of Lebesgue measure, for every $B \in \mathscr{B}(\mathbb{R}^n)$, there exists a compact set $K \subset B$ such that $V_n(B \backslash K)$ is arbitrarily small. Thus, without loss of generality assume A_0 and A_1 are non-empty convex subsets of $\mathbb{R}^n$. For $\epsilon > 0$, let

$$d(x, A) = \inf \{|x - y| : y \in A\}, x \in \mathbb{R}^n,$$

$$A^\epsilon = \{x \in \mathbb{R}^n : d(x, A) \leqslant \epsilon\},$$

$$\phi_A^\epsilon(x) = \min \{\epsilon, d(x, A)\}, \ x \in \mathbb{R}^n.$$

Next, let

$$\tilde{\phi}_\epsilon = \sigma_\theta \min \left\{ \frac{\theta_0}{\sigma_0}, \frac{\theta_1}{\sigma_1} \right\} \phi_{A_\theta^{\epsilon(\theta_0 + \theta_1)}}^\epsilon.$$

Then it is easy to check that

$$\frac{1}{\sigma_\theta} \tilde{\phi}_\epsilon(x_\theta) \leqslant \frac{\theta_0}{\sigma_0} \phi_{A_0}^\epsilon(x_0) + \frac{\theta_1}{\sigma_1} \phi_{A_1}^\epsilon(x_1), \quad x_0, x_1 \in \mathbb{R}^n.$$

From part a) proved above,

$$v_{\sigma_\theta, c_\theta}^{m \tilde{\phi}_\epsilon + F_\theta} (T, x_\theta) \geqslant \left\{ v_{\sigma_0, c_0}^{m \phi_{A_0}^\epsilon + F_0} (T, x_0) \right\}^{\theta_0 \sigma_0 / \sigma_\theta} \left\{ v_{\sigma_1, c_1}^{m \phi_{A_1}^\epsilon + F_1} (T, x_1) \right\}^{\theta_1 \sigma_1 / \sigma_\theta},$$

$$x_0, x_1 \in \mathbb{R}^n, \ t \geqslant 0, \ m \geqslant 1.$$

Notice that on A_j, $\phi_{A_j}^\epsilon = 0$ (on A_θ, $\tilde{\phi}_\epsilon = 0$), while on set $\left\{ \phi_{A_j}^\epsilon > 0 \right\}$, the integrand function tends to zero as $m \to \infty$. Thus, letting first $m \to \infty$, and then $\epsilon \to 0$, we obtain the required inequality. $\square$

Finally, we shall point out that Brunn-Minkowski's inequality is a pure geometric result, not involving randomness. However the probabilistic proof here has not used much special geometric knowledge.

To end this section, we introduce a more direct proof for the Prékopa-Leindler's inequality, which is due to [26].

Theorem 8.30 (Prékopa-Leindler's inequality).   Assume $\lambda \in (0,1)$ and f, g, h are nonnegative and Lebesgue integrable, such that

$$h((1 - \lambda)x + \lambda y) \geqslant f(x)^{1-\lambda} g(y)^{\lambda}, \qquad x, y \in \mathbb{R}^n.$$

Then

$$\int_{\mathbb{R}^n} h(x) \mathrm{d}x \geqslant \left(\int_{\mathbb{R}^n} f(x) \mathrm{d}x \right)^{1-\lambda} \left(\int_{\mathbb{R}^n} g(x) \mathrm{d}x \right)^{\lambda}.$$

Proof a) We first prove the one-dimensional case. By homogeneity, we assume

$$\int_{\mathbb{R}} f = \int_{\mathbb{R}} g = 1.$$

Let $u, v \colon (0,1) \to \mathbb{R}$ be the smallest number such that

$$\int_{-\infty}^{u(r)} f = \int_{-\infty}^{v(r)} g = r.$$

Then u and v are strictly increasing, so they are differential almost everywhere, and

$$f(u(r))u'(r) = g(v(r))v'(r) = 1.$$

Let $w(r) = (1 - \lambda)u(r) + \lambda v(r)$. Then by arithmetic-geometric mean inequality,

$$w'(r) = (1 - \lambda)u'(r) + \lambda v'(r) \geqslant u'(r)^{1-\lambda}v'(r)^{\lambda}$$
$$= f(u(r))^{\lambda-1}g(v(r))^{-\lambda},$$

provided $f(u(r)) \neq 0$ and $g(v(r)) \neq 0$. Therefore

$$\int_{\mathbb{R}} h(x)\mathrm{d}x \geqslant \int_{\text{Range}(w)} h(x)\mathrm{d}x \geqslant \int_0^1 h(w(r))w'(r)\mathrm{d}r$$
$$\geqslant \int_0^1 f(u(r))^{1-\lambda}g(v(r))^{\lambda}f(u(r))^{\lambda-1}g(v(r))^{-\lambda}\mathrm{d}r = 1,$$

where the second inequality comes from the fact that for general increasing function w, we can only get

$$\int_{r_1}^{r_2} w'(r)\mathrm{d}r \leqslant w(r_2) - w(r_1)$$

rather than equality. This corresponds to the case that h is an indicator function of an interval. For general h, we can approximate it by simple functions.

b) For the multi-dimensional case, we use the induction argument. Fix $s \in \mathbb{R}$, define $f_s : \mathbb{R}^{n-1} \to [0, \infty)$ by $f_s(x) = f(x, s)$ for $x \in \mathbb{R}^{n-1}$. Define g_s and h_s similarly. Thus, if $s = (1 - \lambda)s_0 + \lambda s_1$, $s_0, s_1 \in \mathbb{R}$, then from the assumption we have

$$h_s\left((1 - \lambda)x + \lambda y\right) \geqslant f_{s_0}(x)^{1-\lambda} g_{s_1}(y)^{\lambda}, \qquad x, y \in \mathbb{R}^{n-1}.$$

So by induction

$$\int_{\mathbb{R}^{n-1}} h_s \mathrm{d}x \geqslant \left(\int_{\mathbb{R}^{n-1}} f_{s_0} \mathrm{d}x\right)^{1-\lambda} \left(\int_{\mathbb{R}^{n-1}} g_{s_1} \mathrm{d}x\right)^{\lambda}.$$

Regarding the left-hand side as a univariate function of s, and regarding the two terms in the right-hand side as univariate functions of s_0 and s_1, respectively, this becomes the one-dimensional case. Thus it follows from the proof of part a) that

$$\int_{\mathbb{R}^n} h \mathrm{d}x = \int_{\mathbb{R}} \left(\int_{\mathbb{R}^{n-1}} h_s \mathrm{d}x\right) \mathrm{d}s \geqslant \left(\int_{\mathbb{R}^n} f\right)^{1-\lambda} \left(\int_{\mathbb{R}^n} g\right)^{\lambda}. \quad \square$$

The key for the proof of Theorem 8.30 is to introduce the change of variables u and v. With the help of these two functions, we couple the distributions $f\mathrm{d}x$ and $g\mathrm{d}x$ with the uniform distribution. This method is valid for multidimensional Euclid space, however it is much more difficult. The study for the existence, regularity and various applications constitutes an important research topic of the modern partial differential equation and convex geometry.

The Prékopa-Leindler's inequality in Theorem 8.30 is more general than that proved above (below Theorem 8.29), since there we require that f, g and h are bounded and continuous. However, by using suitable approximation procedure, we can prove they are equivalent to each other.

8.10   Supplements and Exercises

(1) (Integral by parts formula) The classical formula is as follows: assume f, g are right-continuous and increasing real functions. Then

$$(fg)\big|_a^b = \int_a^b f(x)\mathrm{d}g(x) + \int_a^b g(x-)\mathrm{d}f(x).$$

Hint: Its proof is not difficult:

$$\big(f(b) - f(a)\big)\big(g(b) - g(a)\big)$$

$$= \iint\limits_{a \leqslant x \leqslant y \leqslant b} \mathrm{d}f(x)\mathrm{d}g(y) + \iint\limits_{a \leqslant y < x \leqslant b} \mathrm{d}f(x)\mathrm{d}g(y)$$

$$= \int_a^b \mathrm{d}g(y) \int_a^y \mathrm{d}f(x) + \int_a^b \mathrm{d}f(x) \int_a^{x-} \mathrm{d}g(y)$$

$$= \int_a^b f(y)\mathrm{d}g(y) - f(a)[g(b) - g(a)]$$

$$+ \int_a^b g(x-)\mathrm{d}f(x) - g(a)[f(b) - f(a)].$$

Try to prove the integration by parts formula in the stochastic case. Given

$$X_t = X_0 + M_t + U_t, \quad Y_t = Y_0 + N_t + V_t, \quad t \geqslant 0,$$

where $(M_t), (N_t) \in \mathcal{M}_c^{loc}$, (U_t) and (V_t) are continuous and bounded variation processes with $U_0 = V_0 = 0$, we have

$$\int_0^T X_t \mathrm{d}Y_t = X_T Y_T - X_0 Y_0 - \int_0^T Y_t \mathrm{d}X_t - \langle M, N \rangle(T), \quad T \geqslant 0.$$

(2) For stopping time τ and $X \in \mathcal{M}_c^2$, let

$X_t^\tau = X_{t \wedge \tau}, \quad t \geqslant 0$ (stop at τ, that is the process stops at time τ).

$^\tau X_t = X_t - X_t^\tau, \quad t \geqslant 0$ (make it to be constant zero up to time τ) .

Prove that X^τ, $^\tau X \in \mathcal{M}_\tau^2$ and

$$\langle X^\tau \rangle(t) = \langle X \rangle(t \wedge \tau), \qquad \langle ^\tau X \rangle(t) = \langle X \rangle(t) - \langle X \rangle(t \wedge \tau), \qquad \text{a.s.}$$

Hint:

a) Assume $t \geqslant s$. Then

$$\mathbb{E}[X_{t \wedge \tau}(X_{t \wedge \tau} - X_t)|\mathscr{F}_s]$$

$$= \mathbb{E}[I_{[\tau \leqslant s]} X_\tau (X_\tau - X_t)|\mathscr{F}_s] + \mathbb{E}[I_{[\tau > s]} X_{t \wedge \tau}(X_{t \wedge \tau} - X_t)|\mathscr{F}_s].$$

The first term on the right-hand side

$$= X_\tau^2 I_{[\tau \leqslant s]} - X_\tau I_{[\tau \leqslant s]} \mathbb{E}[X_t|\mathscr{F}_s]$$

$$= X_\tau (X_\tau - X_t) I_{[\tau \leqslant s]} = X_{s \wedge \tau}(X_{s \wedge \tau} - X_t), \qquad \text{a.s.}$$

The second term on the right-hand side

$$= I_{[\tau > s]} \mathbb{E}[X_{t \wedge \tau}(X_{t \wedge \tau} - X_t)|\mathscr{F}_{s \wedge \tau}]$$

$$= I_{[\tau > s]} \mathbb{E}\big[\mathbb{E}[X_{t \wedge \tau}(X_{t \wedge \tau} - X_t)|\mathscr{F}_{t \wedge \tau}]|\mathscr{F}_{s \wedge \tau}\big] = 0, \qquad \text{a.s.}$$

Thus $(X_{t \wedge \tau}(X_{t \wedge \tau} - X_t), \mathscr{F}_t, \mathbb{P})$ is a martingale.

b)   To calculus $\langle {}^\tau X\rangle(t)$, use

$$(X_t - X_{t\wedge\tau})^2 = X_t^2 - X_{t\wedge\tau}^2 + 2X_{t\wedge\tau}(X_{t\wedge\tau} - X_t).$$

(3) (a) For $X, Y \in \mathcal{M}_c^2((\mathscr{F}_t), \mathbb{P})$ and stopping time τ, prove that

$$\langle X^\tau, Y\rangle = \langle X, Y\rangle^\tau,$$

where $\langle X^\tau, Y\rangle(dt) = I_{[0,\tau)}(t)\langle X, Y\rangle(dt)$.

(b) Let $\mathscr{F}_t^X = \sigma(X_s : 0 \leqslant s \leqslant t), \mathscr{F}_t^Y = \sigma(Y_s : 0 \leqslant s \leqslant t)$. Prove: $X, Y \in \mathcal{M}_c^2((\mathscr{F}_t^X \times \mathscr{F}_t^Y), \mathbb{P})$; $\langle X, Y\rangle$ defined according to $((\mathscr{F}_t^X \times \mathscr{F}_t^Y), \mathbb{P})$ coincides with that defined according to $((\mathscr{F}_t), \mathbb{P})$ probably except a $\mathbb{P}$-null set.

(c) If $\mathscr{F}_T^X$ and $\mathscr{F}_T^Y$ are independent for some $T > 0$, then $\langle X, Y\rangle(t) = 0$, $0 \leqslant t \leqslant T$.

Hint: Since $XY - \langle X, Y\rangle$ is a martingale, so is $(XY)^\tau - \langle X, Y\rangle^\tau = X^\tau Y^\tau - \langle X, Y\rangle^\tau$. To prove $X^\tau Y - \langle X, Y\rangle^\tau$ is a martingale, we need prove only $X^\tau(Y - Y^\tau)$ is a martingale. But when $t > s$,

$$\mathbb{E}\big[X_t^\tau(Y_t - Y_t^\tau)|\mathscr{F}_s\big]$$
$$= \mathbb{E}\big[I_{[\tau\leqslant s]}X_\tau(Y_t - Y_\tau)|\mathscr{F}_s\big] + I_{[\tau>s]}\mathbb{E}\big[X_t^\tau(Y_t - Y_t^\tau)|\mathscr{F}_{s\wedge\tau}\big]$$
$$= I_{[\tau\leqslant s]}X_\tau(Y_s - Y_\tau) + I_{[\tau>s]}\mathbb{E}\big[\mathbb{E}[X_t^\tau(Y_t - Y_t^\tau)|\mathscr{F}_{t\wedge\tau}]|\mathscr{F}_{s\wedge\tau}\big]$$
$$= X_s^\tau(Y_s - Y_s^\tau), \qquad \text{a.s.}$$

(4) Given $X \in \mathcal{M}_c^2$ and stopping times $\sigma \leqslant \tau$, assume $\gamma \in \mathscr{F}_\sigma$ satisfying

$$\mathbb{E}\big[\gamma^2 \cdot \big(\langle X\rangle(T \wedge \tau) - \langle X\rangle(T \wedge \sigma)\big)\big] < \infty.$$

Now let $\alpha = I_{[\sigma,\tau)}\gamma$. Prove that $\alpha \in L_{\ell oc}^2(\langle X\rangle, \mathbb{P})$ and $\alpha_\bullet X$ exists and equals $\gamma\big(X^\tau(\bullet) - X^\sigma(\bullet)\big)$.

Hint: Prove the last assertion only. The proof is divided to three steps.

a) Assume $(Z_t) \in \mathcal{M}_c^2$, $Z^\sigma = 0$. Then

$$\langle Z, Y\rangle^\sigma = 0, \qquad Y \in \mathcal{M}_c^2.$$
$$Z^\sigma = 0 \Rightarrow Z^\sigma Y = 0 \Rightarrow \langle Z^\sigma, Y\rangle = 0 \Rightarrow \langle Z, Y\rangle^\sigma = 0$$
$$\Rightarrow \langle Z, Y\rangle_t = \langle Z, Y\rangle_{t\vee\sigma}.$$

b) If $\gamma \in \mathscr{F}_\sigma$, then $\langle \gamma Z, Y\rangle_t = \langle \gamma Z, Y\rangle_{t\vee\sigma} = \gamma\langle Z, Y\rangle_{t\vee\sigma}$. Indeed,

$$\mathbb{E}[\gamma Z_{t\vee\sigma}Y_{t\vee\sigma} - \gamma\langle Z, Y\rangle_{t\vee\sigma}|\mathscr{F}_{s\vee\sigma}] = \gamma\mathbb{E}[Z_{t\vee\sigma}Y_{t\vee\sigma} - \langle Z, Y\rangle_{t\vee\sigma}|\mathscr{F}_{s\vee\sigma}]$$
$$= \gamma[Z_{s\vee\sigma}Y_{s\vee\sigma} - \langle Z, Y\rangle_{s\vee\sigma}], \quad \text{a.s.}$$

c) Now take $Z = X^\tau - X^\sigma$. Then

$$\langle \gamma(X^\tau - X^\sigma), Y \rangle = \gamma \langle X^\tau - X^\sigma, Y \rangle = \gamma(\langle X^\tau, Y \rangle - \langle X^\sigma, Y \rangle)$$
$$= \gamma(\langle X, Y \rangle^\tau - \langle X, Y \rangle^\sigma).$$

Thus

$$\langle \gamma(X^\tau - X^\sigma), Y \rangle(dt) = \gamma(\langle X, Y \rangle^\tau(dt) - \langle X, Y \rangle^\sigma(dt))$$
$$= \gamma \Big(I_{[0,\tau)} \langle X, Y \rangle(dt) - I_{[0,\sigma)} \langle X, Y \rangle(dt) \Big)$$
$$= \gamma I_{[\sigma,\tau)} \langle X, Y \rangle(dt).$$

(5) Assume $X \in \mathcal{M}_c^2$.

(a) The localization of integrand is equivalent to the localization of the integral. Assume $\sigma \leqslant \tau$ are stopping times, $\alpha \in L^2_{\ell oc}(\langle X \rangle, \mathbb{P})$. Prove that $I_{[\sigma,\tau)} \alpha_t \in L^2_{\ell oc}(\langle X \rangle, \mathbb{P})$ and

$$\int_{T \wedge \sigma}^{T \wedge \tau} \alpha_t dX_t \equiv \int_0^T I_{[\sigma,\tau)} \alpha_t dX_t = \int_0^{T \wedge \tau} \alpha_t dX_t - \int_0^{T \wedge \sigma} \alpha_t dX_t, \text{ a.s.}$$

(b) (Differential of stochastic integral)   Assume $\beta \in L^2_{\ell oc}(\langle X \rangle, \mathbb{P})$ and $\alpha \in L^2_{\ell oc}(\beta^2 \langle X \rangle, \mathbb{P})$. Prove:

$$\int_0^T \alpha_t d \Big(\int_0^t \beta_s dX_s \Big) = \int_0^T \alpha_s \beta_s dX_s, \qquad \text{a.s.}$$

Hint: The equality in (a) only uses the linearity of stochastic integral.

$$\int_0^T I_{[\sigma,\tau)}(t) \alpha_t dX_t = \int_0^T \Big(I_{[0,\tau)(t)} \alpha_t - I_{[0,\sigma)} \alpha(t) \Big) dX_t$$
$$= \int_0^T I_{[0,\tau)(t)} \alpha_t dX_t - \int_0^T I_{[0,\sigma)} \alpha_t dX_t$$
$$= \int_0^{T \wedge \tau} \alpha_t dX_t - \int_0^{T \wedge \tau} \alpha_t dX_t.$$

Next for (b), note that for every $Y \in \mathcal{M}_c^2$,

$$\Big\langle\!\!\Big\langle \int_0^\bullet \alpha_u d \Big(\int_0^u \beta_s dX_s \Big), Y \Big\rangle\!\!\Big\rangle(dt) = \alpha_t \Big\langle\!\!\Big\langle \int_0^\bullet \beta_s dX_s, Y \Big\rangle\!\!\Big\rangle(dt)$$
$$= \alpha_t \beta_t \langle X, Y \rangle(dt) = \Big\langle\!\!\Big\langle \int_0^\bullet \alpha_u \beta_u dX_u, Y \Big\rangle\!\!\Big\rangle(dt).$$

Therefore the assertion follows from the uniqueness.

(6) Assume X and σ are as above. Next, assume $Y \in (\mathscr{M}_c^2)^m$ and
$$\tau : [0, \infty) \times \Omega \to \mathbb{R}^N \otimes \mathbb{R}^m$$
satisfy the similar conditions. Prove that

(a) $\int_0^{\cdot} [\sigma, \tau] \, d \begin{bmatrix} X \\ Y \end{bmatrix} = \int_0^{\cdot} \sigma dX + \int_0^{\cdot} \tau dY$;

(b) $\left\langle\!\left\langle \int_0^{\cdot} \sigma dX, \int_0^{\cdot} \tau dY \right\rangle\!\right\rangle = \sigma(t) \langle\!\langle X, Y \rangle\!\rangle(dt) \tau(t)^*$, where
$$\langle\!\langle X, Y \rangle\!\rangle = (\langle X^i, Y^j \rangle : 1 \leqslant i \leqslant d, \; 1 \leqslant j \leqslant m);$$

(c) $\int_0^{\cdot} \sigma dX$ is the unique $Y \in (\mathscr{M}_c^2)^N$ such that $Y(0) = 0$ and
$$\left\langle\!\left\langle \begin{bmatrix} X \\ Y \end{bmatrix}, \begin{bmatrix} X \\ Y \end{bmatrix} \right\rangle\!\right\rangle(dt) = \begin{bmatrix} I \\ \sigma(t) \end{bmatrix} \langle\!\langle X, X \rangle\!\rangle(dt) \, [I, \sigma(t)^*], \quad \text{a.s.}$$
where I is the identity matrix.

(d) Assume $\tau : [0, \infty) \times \Omega \to \mathbb{R}^M \otimes \mathbb{R}^N$ is progressively measurable, σ and $\tau\sigma$ satisfy the above condition, and
$$\mathbb{E}\left[\int_0^T \mathrm{Tr}\big(\sigma(t) \langle\!\langle X, X \rangle\!\rangle(dt) \sigma^*(t) \big) \right] < \infty, \quad T > 0.$$
Then
$$\int_0^{\cdot} \tau(t) d\left(\int_0^t \sigma dX \right) = \int_0^{\cdot} \tau\sigma dX, \quad \text{a.s.}$$

Hint:

(i) By definition,
$$(\theta, \sigma_{\bullet} X)_{\mathbb{R}^N} = \int_0^{\cdot} (\sigma^* \theta) dX, \quad \theta \in \mathbb{R}^N.$$
Thus, for $\theta \in \mathbb{R}^N$,
$$\left(\theta, \int_0^{\cdot} [\sigma, \tau] d \begin{bmatrix} X \\ Y \end{bmatrix} \right)_{\mathbb{R}^N} = \int_0^{\cdot} [\sigma, \tau]^* \theta \, d \begin{bmatrix} X \\ Y \end{bmatrix}$$
$$= \int_0^{\cdot} \begin{bmatrix} \sigma^* \\ \tau^* \end{bmatrix} \theta \, d \begin{bmatrix} X \\ Y \end{bmatrix} = \int_0^{\cdot} \begin{bmatrix} \sigma^* \theta \\ \tau^* \theta \end{bmatrix} d \begin{bmatrix} X \\ Y \end{bmatrix},$$
$$\left(\theta, \int_0^{\cdot} \sigma dX + \int_0^{\cdot} \tau^* \theta dY \right) = \int_0^{\cdot} \sigma^* \theta dX + \int_0^{\cdot} \tau^* \theta dY.$$
Therefore, we need only prove
$$\int_0^{\cdot} \begin{bmatrix} \sigma^* \theta \\ \tau^* \theta \end{bmatrix} d \begin{bmatrix} X \\ Y \end{bmatrix} = \int_0^{\cdot} \sigma^* \theta dX + \int_0^{\cdot} \tau^* \theta dY.$$
This holds in the special case that $\sigma^* \theta$ and $\tau^* \theta$ belong to $(L_{\ell oc}^2)^N$. Then by taking limit, we pass to the general case.

(ii) Similar to the proof for $\langle \theta_\bullet X, \eta_\bullet Y \rangle$, we can prove:

$$\langle \theta_\bullet X, \eta_\bullet Y \rangle = \theta \langle\!\langle X, Y \rangle\!\rangle \eta.$$

Now, for each $\theta \in \mathbb{R}^N$,

$$\left\langle \int_0^\bullet \sigma^* \theta \mathrm{d}X, \int_0^\bullet \tau^* \theta \mathrm{d}Y \right\rangle (\mathrm{d}t)$$
$$= (\sigma^* \theta)^*(t) \langle\!\langle X, Y \rangle\!\rangle (\mathrm{d}t)(\tau^* \theta)(t) = \theta^* \sigma(t) \langle\!\langle X, Y \rangle\!\rangle (\mathrm{d}t) \tau^*(t) \theta.$$

Take θ to be a basis in $\mathbb{R}^N$, and use $\langle \theta, \int_0^\bullet \sigma \mathrm{d}X \rangle$, we can compute out each component.

(iii) Prove first that $Y = \int_0^\bullet \sigma \mathrm{d}X$ satisfies the equality. Since

$$\begin{bmatrix} X \\ Y \end{bmatrix} = \begin{bmatrix} X \\ \int_s^\bullet \sigma \mathrm{d}X \end{bmatrix} = \int_0^\bullet \begin{bmatrix} I \\ \sigma \end{bmatrix} \mathrm{d}\begin{bmatrix} X \\ X \end{bmatrix},$$

the equality follows from (ii). Next prove the uniqueness. From the above equation, we have

$$\langle\!\langle Y, X \rangle\!\rangle (\mathrm{d}t) = \sigma(t) \langle\!\langle X, X \rangle\!\rangle (\mathrm{d}t).$$

Rewrite it in the component form

$$\langle Y^i, X^j \rangle (\mathrm{d}t) = \sum_k \sigma_{ik}(t) \langle X^k, X^j \rangle (\mathrm{d}t).$$

But $\langle Y^i, \int_0^\bullet \eta_j \mathrm{d}X^j \rangle (\mathrm{d}t) = \eta_j(t) \langle Y^i, X^j \rangle (\mathrm{d}t)$. Therefore

$$\left\langle Y^i, \int_0^\bullet \eta \mathrm{d}X \right\rangle (\mathrm{d}t) = \sigma_{i \bullet}(t) \langle\!\langle X, X \rangle\!\rangle (\mathrm{d}t) \eta(t).$$

Then using the uniqueness in the previous section, we obtain

$$Y^i = \int_0^\bullet \sigma^*_{\bullet i}(t) \mathrm{d}X_t.$$

Moreover

$$(\theta, Y)_{\mathbb{R}^N} = \sum_{i=1}^N \theta_i Y^i = \int_0^\bullet \sum_{i=1}^N \sigma^*_{\bullet i}(t) \theta_i \mathrm{d}X_t = \int_0^\bullet (\sigma^*(t) \theta) \mathrm{d}X_t.$$

Thus $Y = \int_0^\bullet \sigma(t) \mathrm{d}X_t$.

(7) Prove the multidimensional Girsanov theorem (the case of $N > 1$ in Theorem 8.27).

Notes

In this part, we introduce some books for further reading and explain the materials used in our book. Most of the topics in this book are certainly classical, we are not able to point out the exact originality of the materials due to the lack of the study on the developing history. For the new materials, we wish to mention the original references.

1. The books for further reading:

- For the Chinese books on fundamentals of stochastic processes, refer to [66] (or [67]), [36], [30], [47], [53], [54], [28], [38]. For the English books, refer to [44], [27], [50], [42], [25]. From these books, the readers can not only find more richer contents or different arguments for the same topics in our book, but also find more extensive topics excluded in our book. And the reader can find more books that the stochastic processes are applied to engineering, public service, medicine, ecology, finance, computer science and so on.
- For Markov chains and general Markov jump processes, refer to [68], [33], [34], [35], [73], [5], [10], [19], [1], [48].
- For modern theory on stochastic analysis, refer to [70], [31], [71], [39], [63].

The readers can at least browse through some chapters in these books, to appreciate some scenes of the new developments of modern stochastic processes. For those who want to do research on stochastic processes, we suggest they should read intensively on one or two such books. In the following, we will present some publications for further reading about each chapter in our book.

2. Notes.

We begin with the first part.

- §1.1 is taken from [6], see also [11]. For the application of the search engine in Remark 1.9, refer to the survey article [45].
- §1.2 and §1.3 are almost taken from [66], while the coupling method used in the proof of ergodicity is taken from [2] and [53].
- §1.4 is taken from [5] and [10]. The systematic study on the non-negative minimal solution comes from [33].
- §1.5 comes from [43].
- §2.1 is basically classical, the original references are [5], [10], [33], [53] or the books listed above for Markov chains. The sufficient part of Theorem 2.9 can be found in [5] or [10], and the necessary part was mainly proved by [61]. For this theorem and related results, refer to the survey article [12], where an alternative proof for the necessity is included.
- §2.2: Theorem 2.13 is taken from [5]. The sufficient part in Theorem 2.15 is due to [56], while the necessary part is due to [53]. Tweedie's result comes from [64]. For ergodicity, the complete argument comes from [10].
- §2.3: The single birth processes without absorbing states were studied in [5], which contains some mistakes for the single processes with absorbing states.
- §2.4: The branching processes are discussed in many textbooks, but the extended branching processes appeared much late. The materials are taken from [18].
- §3.1: The materials in this section are taken from [17]. The contents in this section can be found in [5] or [10].
- §3.2: The results in this section are accomplished by Mu-Fa Chen in a series of papers. The estimate of spectral gap and related topics are an active research direction today. For more details, refer to [11], [14] and [65].
- §3.3: This section is written for the beginners, and is published in English for the first time.
- §4.1 and §4.2 are taken from [66]. It is first published in English the detailed study on the definitions of the stopping time T and σ-algebra $\mathscr{F}_T$.
- §4.3: The problem of the optimal stopping is one of the important branches in probability theory, which has many important implications. Refer to [40].

Let us go to the second part.

- §5.1–§5.3 are taken from [70], see also [31]. The first example in §5.6 is taken from [20], and the second one is a typical application of martingale theory [63], whose advantage is that this argument can also deal with the time-inhomogeneous Markov processes. This is the theme of the doctoral dissertation [77].
- §6.1–§6.4 are basically taken from [66].
- §7.1–§7.3 are taken from [50] and [29].
- §7.4: Theorems 7.9–7.11 are taken from [39]. Theorem 7.12 comes from [16], where the analytic proof is taken from [8], and the final assertion is an improvement of [9].
- §8.1–§8.8 are taken from [62].
- §8.9: The three mathematical tools should be included in every book on stochastic analysis. Although Proposition 8.26 is simple, it is the starting point for the research on the stochastically conformal theory. See the survey paper [69]. Theorem 8.29 is due to [3]. For Theorem 8.30 and more geometric applications, refer to the survey paper [26].
- Finally, we would like to acknowledge the probabilists whose names may or may not appear in this book for their contributions to the subject, and also thank deeply those authors for their valuable contributions to the book through their publications.

Bibliography

[1] Anderson, W. J. *Continuous-Time Markov Chains.* Springer Series in Statistics, 1991.

[2] Billingsley, P. *Probability and Measure.* 3rd ed. Wiley, 1995.

[3] Borell, C. *Diffusion equations and geometric inequalities.* Potential Anal., 12, 49-71, 2000.

[4] Chen, Mu-Fa. *Coupling of jump processes.* Acta Math. Sinica, New Series, 2(2), 123-136, 1986.

[5] Chen, Mu-Fa. *Jump Processes and Particle Systems* (in Chinese). Beijing Normal University Press, 1986.

[6] Chen, Mu-Fa. *Stochastic models in economic optimization* (in Chinese). Chinese Applied Probability and Statistics, (I): 8(3), 289-294; (II): 8(4), 374-377, 1992.

[7] Chen, Mu-Fa. *Single birth processes.* Chin. Ann. of Math., 20B(1), 77-82, 1999.

[8] Chen, Mu-Fa. *Analytic proof of dual variational formula for the first eigenvalue in dimension one.* Sci. Sin. (A), 29(4), 327-336, 1999.

[9] Chen, Mu-Fa. *Variational formulas and approximation theorems for the first eigenvalue.* Sci. China (A), 44(4):409-418, 2001.

[10] Chen, Mu-Fa. *From Markov Chains to Non-equilibrium Particle Systems.* 2nd ed. World Scientific, 2004.

[11] Chen, Mu-Fa. *Eigenvalues, Inequalities, and Ergodic Theory.* Springer, 2005.

[12] Chen, Mu-Fa. *Practical criterion for uniqueness of Q-processes.* Chinese J. Appl. Prob. Stat. 31(2), 213-224, 2015.

[13] Chen, Mu-Fa. *Efficient initials for computing the maximal eigenpair.* Front. Math. China, 11(6), 1379-1418, 2016.

[14] Chen, Mu-Fa. *Unified speed estimation of various stabilities* (in Chinese). Chinese J. Prob. Stat., 32(1), 1-22, 2016.

[15] Chen, Mu-Fa, Li, Yong. *Stochastic model of economic optimization* (in Chinese). J. Beijing Normal Univ., 30(2), 185-194, 1994.

[16] Chen, Mu-Fa, Wang, Feng-Yu. *Estimation of spectral gap for elliptic operators*, Tran. Amer. Math. Soc., 349, 1239-1267, 1997.

[17] Chen, Mu-Fa, Wang, Pei-Zhuang, Hou, Zhen-Ting, Guo, Qing-Feng, Qian, Min, Qian, Minpin, Gong, Guanglu. *Reversible Markov Processes* (in Chinese). Hunan Scientific Press, 1979.

[18] Chen, Rong-Rong. *An extended class of time-continuous branching processes*. J. Appl. Prob., 34(1), 14-23, 1997.

[19] Chung, Kai Lai. *Markov Chains with Stationary Transition Probabilities*. 2nd ed. Springer, 1967.

[20] Chung, Kai Lai. *A Course on Probability Theory*. 3rd ed. New York, Academic Press, 2001.

[21] Doyle, P. G., Snell, J. L. *Random walks and electric networks*. The Mathematical Association of America, 1984.

[22] Fang, Shi-Zhan, Zhang, Tu-Sheng. *Stochastic differential equations with non-Lipschitz coefficients: path-wise uniqueness and non-explosion*. J. Funct. Anal., 213(2), 440-465, 2004.

[23] Feller, W. *On the integro-differential equations of pure discontinuous Markov processes*. Tran. Amer. Math. Soc., 48, 488-515, 1940.

[24] Feller, W. *On boundaries and lateral conditions for Kolmogorof differential equations*. Ann. Math., 65, 527-570, 1957.

[25] Friedman, A. *Stochastic Differential Equations and Applications, Vol. I, II*. Academic Press, 1975-1976.

[26] Gardner, R. J. *The Brunn-Minkowski inequality*. Bull. (New Ser.) Amer. Math. Soc., 39(3), 355-405, 2002.

[27] Gihman, I. I., Skorohod, A. V. *The Theory of Stochastic Processes, Vol. I, II, III*. Springer, 1979.

[28] Gong, Guanglu. *Introduction to Stochastic Differential Equations* (in Chinese). 2nd ed. Peking University Press, 2000.

[29] Grimmett, G. R., Stirzaker, D. R. *Probability and Random processes*. 3rd ed. Oxford University Press, 2001.

[30] He, Shengwu. *Stochastic Processes* (in Chinese). East China Normal University Press, 1989.

[31] He, Shengwu, Wang, Jiagang, Yan, Jiaan. *Semi-Martingales and Stochastic Analysis* (in Chinese). Science Press, 1995.

[32] Hennion, H. *Limit theorems for products of positive matrices*. Ann. Probab., 25(4), 1545-1587.

[33] Hou, Zhenting, Guo, Qingfeng. *Homogeneous denumerable Markov Processes*. Springer, 1988.

[34] Hou, Zhenting, Zou, Jiezhong, Zhang, Hanjun, Liu, Zaiming, Xiao, Guoneng, Chen, Anyue, Fei, Zhilin. *Q-Matrix Problems for Markov processes* (in Chinese). Hunan Scientific Process, 1994.

[35] Hou, Zhenting, Liu, Zaiming, Zhang, Hanjun, Li, Junping, Zou, Jiezhong, Yuan, Chenghui. *Birth and Death Processes* (in Chinese). Hunan Scientific Press, 2000.

[36] Hu, Di-He. *Stochastic Processes* (in Chinese). Wuhan University Press, 2000.

[37] Hua, Luo-Geng. *The mathematical theory of global optimization on planned economy, parts II and III* (in Chinese). Kexue Tongbao, 13, 769-772, 1984.

[38] Huang, Zhi-Yuan. *Basic Stochastic Analysis* (in Chinese). Science Press, 2001.

[39] Iketa, N, Watanabe, S. *Stochastic Differential Equations and Diffusion Processes*. North-Holland, 1981.

[40] Jin, Zhiming. *Theory of Optimization and Its Applications* (in Chinese). Defense Scientific University Press, 1995.

[41] Kallenberg, O. *Foundation of Modern Probability*. 2nd ed. Springer-Verlag, 2001.

[42] Karatzas, I., Shreve, S. E. *Brownian Motion and Stochastic Calculus*. Springer, 1991.

[43] Karlin, S., Taylor, H. M. *A First Course in Stochastic Processes*. 2nd ed. Academic Press, 1975.

[44] Karlin, S., Taylor, H. M. *A Second Course in Stochastic Processes*. Academic Press, 1981.

[45] Langville, A. N., Meyer, C. D. *A survey of eigenvector method for web information retrieval*. SIAM Review, 47(1), 135-161, 2005.

[46] Langville, A. N., Meyer, C. D. *Google's PageRank and Beyond: The Science of Search Engine Rankings*. Princeton University Press, 2006.

[47] Li, Zhangnan, Wu, Rong. *Texts on Stochastic Processes*. Higher Education Press, 1987.

[48] Meyn, S. P., Tweedie, R. L. *Markov Chains and Stochastic Stability*. 2nd ed. Springer, 2009.

[49] Neveau, J. *Mathematical Foundations of the Calculus of Probability*. Holden-Day, 1965.

[50] Øksendal, B. K. *Stochastic Differential Equation*, 6th ed. Springer, 2003.

[51] Oguntuase, J. *On an inequality of Gronwall*. J. Inequal. Pure and Appl. Math., 2(1), Art.9, 2001.

[52] Petrov, V. *Limit Theorems of Probability Theory, Sequence of Independent Random Variables*. Clarendon Press, 1995.

[53] Qian, Minpin. *Introduction to Stochastic Processes* (in Chinese). Peking University Press, 1990.

[54] Qian Minpin, Gong, Guanglu. *Applied Stochastic Processes* (in Chinese). Peking University Press, 1998.

[55] Reuter, G. E. H. *Denumerable Markov processes*. Acta Math., 97, 1-46, 1957.

[56] Reuter, G. E. H. *Competition processes*. Fourth Berkeley Symposium on Mathematical Statistics and Probability, Vol. 2, 421-430, 1961.

[57] Revuz, D, Yor, M. *Continuous martingales and Brownian motion*. 3rd ed. Springer, 2005.

[58] Rosenblatt, M. *Random Processes*. 2nd ed. Springer, 1974.

[59] Shiryayev, A. N. *Probability*. 2nd ed. Springer-Verlag, 1996.

[60] Sinclair, A. *Algorithms for Random Generation and Counting: a Markov chain approach*. Birkhäuser, 1993.

[61] Spieksma, F. M. *Countable state Markov processes: non-explosiveness and moment function*. Probab. Eng. and Inform. Sci. 29(4), 623-637, 2015.

[62] Stroock, D W. *Lecture on Stochastic Analysis: Diffusion Theory*. Cambridge University Press, 1987.

[63] Stroock, D. W., Varadhan, S. R. S. *Multidimensional Diffusion Processes*. Reprint of 1997 edition. Springer, 2006.

[64] Tweedie, R. L. *Criteria for ergodicity, exponential ergodicity and strong ergodicity of Markov Processes*. J. Appl. Prob. 18, 122-130, 1981.

[65] Wang, Feng-Yu. *Functional Inequalities, Markov Processes, and Spectral Theory*. Science Press, 2004.

[66] Wang, Zikun. *Stochastic Processes*. Science Press, 1965.

[67] Wang, Zikun. *General Stochastic Processes (Vol I)*. Beijing Normal University Press, 1996.

[68] Wang, Zikun, Yang, Xiangqun. *Birth-Death Processes and Markov Chains*. 2nd Ed. Science Press, 2005.

[69] Werner, W. *Conformal restriction and related questions*. Probability Surveys, Vol. 2, 145-190, 2005.

[70] Yan, Jiaan. *Introduction to Martingales and Stochastic Integration* (in Chinese). Shanghai Scientific Press, 1981.

[71] Yan, Jiaan, Peng, Shige, Fang, Shizhan, Wu, Liming. *Topics on Stochastic Analysis* (in Chinese). Science Press, 1997.

[72] Yan, Shi-Jian, Wang, Xiuxiang, Liu, Xiufang. *Foundation of Probability Theory*. Second edition. Science, 2009.

[73] Yang, Xiangqun. *The Construction Theory of Denumerable Markov Processes*. Wiley, 1990.

[74] Yosida, K. *Functional Analysis*. 6th ed. Springer-Verlag, 1980.

[75] Zhang, Gongqing, Lin, Yuanqu. *Lectures on Functional Analysis* (in Chinese). Vol. I, Peking University Press, 1990.

[76] Zhang, Gongqing, Guo, Maozheng. *Lectures on Functional Analysis* (in Chinese). Vol. II, Peking University Press, 1990.

[77] Zheng, Junli. *Phase Transition of Ising Models on Lattice Fractals and Martingale Methods for q-Processes* (in Chinese). Doctoral Dissertion of Beijing Normal University, 1993.

Index